Big Data Analytics

Saumyadipta Pyne · B.L.S. Prakasa Rao
S.B. Rao
Editors

Big Data Analytics

Methods and Applications

Editors
Saumyadipta Pyne
Indian Institute of Public Health
Hyderabad
India

B.L.S. Prakasa Rao
CRRao AIMSCS
University of Hyderabad Campus
Hyderabad
India

S.B. Rao
CRRao AIMSCS
University of Hyderabad Campus
Hyderabad
India

ISBN 978-81-322-3871-3   ISBN 978-81-322-3628-3   (eBook)
DOI 10.1007/978-81-322-3628-3

This Springer imprint is published by Springer Nature
The registered company is Springer (India) Pvt. Ltd.
The registered company address is: 7th Floor, Vijaya Building, 17 Barakhamba Road, New Delhi 110 001, India

Foreword

Big data is transforming the traditional ways of handling data to make sense of the world from which it is collected. Statisticians, for instance, are used to developing methods for analysis of data collected for a specific purpose in a planned way. Sample surveys and design of experiments are typical examples.

Big data, in contrast, refers to massive amounts of very high dimensional and even unstructured data which are continuously produced and stored with much cheaper cost than they are used to be. High dimensionality combined with large sample size creates unprecedented issues such as heavy computational cost and algorithmic instability.

The massive samples in big data are typically aggregated from multiple sources at different time points using different technologies. This can create issues of heterogeneity, experimental variations, and statistical biases, and would therefore require the researchers and practitioners to develop more adaptive and robust procedures.

Toward this, I am extremely happy to see in this title not just a compilation of chapters written by international experts who work in diverse disciplines involving Big Data, but also a rare combination, within a single volume, of cutting-edge work in methodology, applications, architectures, benchmarks, and data standards.

I am certain that the title, edited by three distinguished experts in their fields, will inform and engage the mind of the reader while exploring an exciting new territory in science and technology.

Calyampudi Radhakrishna Rao
C.R. Rao Advanced Institute of Mathematics,
Statistics and Computer Science,
Hyderabad, India

Preface

The emergence of the field of Big Data Analytics has prompted the practitioners and leaders in academia, industry, and governments across the world to address and decide on different issues in an increasingly data-driven manner. Yet, often Big Data could be too complex to be handled by traditional analytical frameworks. The varied collection of themes covered in this title introduces the reader to the richness of the emerging field of Big Data Analytics in terms of both technical methods as well as useful applications.

The idea of this title originated when we were organizing the "Statistics 2013, International Conference on Socio-Economic Challenges and Sustainable Solutions (STAT2013)" at the C.R. Rao Advanced Institute of Mathematics, Statistics and Computer Science (AIMSCS) in Hyderabad to mark the "International Year of Statistics" in December 2013. As the convener, Prof. Saumyadipta Pyne organized a special session dedicated to lectures by several international experts working on large data problems, which ended with a panel discussion on the research challenges and directions in this area. Statisticians, computer scientists, and data analysts from academia, industry and government administration participated in a lively exchange.

Following the success of that event, we felt the need to bring together a collection of chapters written by Big Data experts in the form of a title that can combine new algorithmic methods, Big Data benchmarks, and various relevant applications from this rapidly emerging area of interdisciplinary scientific pursuit. The present title combines some of the key technical aspects with case studies and domain applications, which makes the materials more accessible to the readers. In fact, when Prof. Pyne taught his materials in a Master's course on "Big and High-dimensional Data Analytics" at the University of Hyderabad in 2013 and 2014, it was well-received.

We thank all the authors of the chapters for their valuable contributions to this title. Also, We sincerely thank all the reviewers for their valuable time and detailed comments. We also thank Prof. C.R. Rao for writing the foreword to the title.

Hyderabad, India   Saumyadipta Pyne
June 2016   B.L.S. Prakasa Rao
                                                                                      S.B. Rao

Contents

About the Editors

Saumyadipta Pyne is Professor at the Public Health Foundation of India, at the Indian Institute of Public Health, Hyderabad, India. Formerly, he was P.C. Mahalanobis Chair Professor and head of Bioinformatics at the C.R. Rao Advanced Institute of Mathematics, Statistics and Computer Science. He is also Ramalingaswami Fellow of Department of Biotechnology, the Government of India, and the founder chairman of the Computer Society of India's Special Interest Group on Big Data Analytics. Professor Pyne has promoted research and training in Big Data Analytics, globally, including as the workshop co-chair of IEEE Big Data in 2014 and 2015 held in the U.S.A. His research interests include Big Data problems in life sciences and health informatics, computational statistics and high-dimensional data modeling.

B.L.S. Prakasa Rao is the Ramanujan Chair Professor at the C.R. Rao Advanced Institute of Mathematics, Statistics and Computer Science, Hyderabad, India. Formerly, he was director at the Indian Statistical Institute, Kolkata, and the Homi Bhabha Chair Professor at the University of Hyderabad. He is a Bhatnagar awardee from the Government of India, fellow of all the three science academies in India, fellow of Institute of Mathematical Statistics, U.S.A., and a recipient of the national award in statistics in memory of P.V. Sukhatme from the Government of India. He has also received the Outstanding Alumni award from Michigan State University. With over 240 papers published in several national and international journals of repute, Prof. Prakasa Rao is the author or editor of 13 books, and member of the editorial boards of several national and international journals. He was, most recently, the editor-in-chief for journals—*Sankhya* A and *Sankhya* B. His research interests include asymptotic theory of statistical inference, limit theorems in probability theory and inference for stochastic processes.

S.B. Rao was formerly director of the Indian Statistical Institute, Kolkata, and director of the C.R. Rao Advanced Institute of Mathematics, Statistics and Computer Science, Hyderabad. His research interests include theory and algorithms in graph theory, networks and discrete mathematics with applications in social, biological, natural and computer sciences. Professor S.B. Rao has 45 years of teaching and research experience in various academic institutes in India and abroad. He has published about 90 research papers in several national and international journals of repute, and was editor of 11 books and proceedings. He wrote a paper jointly with the legendary Hungarian mathematician Paul Erdos, thus making his "Erdos Number" 1.

Big Data Analytics: Views from Statistical and Computational Perspectives

Saumyadipta Pyne, B.L.S. Prakasa Rao and S.B. Rao

Abstract Without any doubt, the most discussed current trend in computer science and statistics is BIG DATA. Different people think of different things when they hear about big data. For the statistician, the issues are how to get usable information out of datasets that are too huge and complex for many of the traditional or classical methods to handle. For the computer scientist, big data poses problems of data storage and management, communication, and computation. For the citizen, big data brings up questions of privacy and confidentiality. This introductory chapter touches some key aspects of big data and its analysis. Far from being an exhaustive overview of this fast emerging field, this is a discussion on statistical and computational views that the authors owe to many researchers, organizations, and online sources.

1 Some Unique Characteristics of Big Data

Big data exhibits a range of characteristics that appears to be unusual when compared to traditional datasets. Traditionally, datasets were generated upon conscious and careful planning. Field experts or laboratory experimenters typically spend considerable time, energy, and resources to produce data through planned surveys or designed experiments. However, the world of big data is often nourished by dynamic sources such as intense networks of customers, clients, and companies, and thus there is an automatic flow of data that is always available for analysis. This almost voluntary generation of data can bring to the fore not only such obvious issues as data volume, velocity, and variety but also data veracity, individual privacy, and indeed,

S. Pyne (✉)
Indian Institute of Public Health, Hyderabad, India
e-mail: spyne@iiphh.org

B.L.S. Prakasa Rao · S.B. Rao
C.R. Rao Advanced Institute of Mathematics, Statistics and Computer Science,
Hyderabad, India
e-mail: blsprao@gmail.com

S.B. Rao
e-mail: siddanib@yahoo.co.in

S. Pyne et al. (eds.), *Big Data Analytics*, DOI 10.1007/978-81-322-3628-3_1

ethics. If data points appear without anticipation or rigor of experimental design, then their incorporation in tasks like fitting a suitable statistical model or making a prediction with a required level of confidence, which may depend on certain assumptions about the data, can be challenging. On the other hand, the spontaneous nature of such real-time pro-active data generation can help us to capture complex, dynamic phenomena and enable data-driven decision-making provided we harness that ability in a cautious and robust manner. For instance, popular Google search queries could be used to predict the time of onset of a flu outbreak days earlier than what is possible by analysis of clinical reports; yet an accurate estimation of the severity of the outbreak may not be as straightforward [1]. A big data-generating mechanism may provide the desired statistical power, but the same may also be the source of some its limitations.

Another curious aspect of big data is its potential of being used in unintended manner in analytics. Often big data (e.g., phone records) could be used for the type of analysis (say, urban planning) that is quite unrelated to the original purpose of its generation, especially if the purpose is integration or triangulation of diverse types of data, including auxiliary data that may be publicly available. If a direct survey of a society's state of well-being is not possible, then big data approaches can still provide indirect but valuable insights into the society's socio-economic indicators, say, via people's cell phone usage data, or their social networking patterns, or satellite images of the region's energy consumption or the resulting environmental pollution, and so on. Not only can such unintended usage of data lead to genuine concerns about individual privacy and data confidentiality, but it also raises questions regarding enforcement of ethics on the practice of data analytics.

Yet another unusual aspect that sometimes makes big data what it is is the rationale that if the generation costs are low, then one might as well generate data on as many samples and as many variables as possible. Indeed, less deliberation and lack of parsimonious design can mark such "glut" of data generation. The relevance of many of the numerous variables included in many big datasets seems debatable, especially since the outcome of interest, which can be used to determine the relevance of a given predictor variable, may not always be known during data collection. The actual explanatory relevance of many measured variables to the eventual response may be limited (so-called "variable sparsity"), thereby adding a layer of complexity to the task of analytics beyond more common issues such as data quality, missing data, and spurious correlations among the variables.

This brings us to the issues of high variety and high dimensionality of big data. Indeed, going beyond structured data, which are "structured" in terms of variables, samples, blocks, etc., and appear neatly recorded in spreadsheets and tables resulting from traditional data collection procedures, increasingly, a number of sources of unstructured data are becoming popular—text, maps, images, audio, video, news feeds, click streams, signals, and so on. While the extraction of the essential features can impose certain structure on it, unstructured data nonetheless raises concerns regarding adoption of generally acceptable data standards, reliable annotation of metadata, and finally, robust data modeling. Notably, there exists an array of powerful tools that is used for extraction of features from unstructured data, which allows combined modeling of structured and unstructured data.

Let us assume a generic dataset to be a $n \times p$ matrix. While we often refer to big data with respect to the number of data points or samples therein (denoted above by n), its high data volume could also be due to the large number of variables (denoted by p) that are measured for each sample in a dataset. A high-dimensional or "big p" dataset (say, in the field of genomics) can contain measurements of tens of thousands of variables (e.g., genes or genomic loci) for each sample. Increasingly, large values of both p and n are presenting practical challenges to statisticians and computer scientists alike. High dimensionality, i.e., big p relative to low sample size or small n, of a given dataset can lead to violation of key assumptions that must be satisfied for certain common tests of hypotheses to be applicable on such data. In fact, some domains of big data such as finance or health do even produce infinite dimensional functional data, which are observed not as points but functions, such as growth curves, online auction bidding trends, etc.

Perhaps the most intractable characteristic of big data is its potentially relentless generation. Owing to automation of many scientific and industrial processes, it is increasingly feasible, sometimes with little or no cost, to continuously generate different types of data at high velocity, e.g., streams of measurements from astronomical observations, round-the-clock media, medical sensors, environmental monitoring, and many "big science" projects. Naturally, if streamed out, data can rapidly gain high volume as well as need high storage and computing capacity. Data in motion can neither be archived in bounded storage nor held beyond a small, fixed period of time. Further, it is difficult to analyze such data to arbitrary precision by standard iterative algorithms used for optimal modeling or prediction. Other sources of intractability include large graph data that can store, as network edges, static or dynamic information on an enormous number of relationships that may exist among individual nodes such as interconnected devices (the Internet of Things), users (social networks), components (complex systems), autonomous agents (contact networks), etc. To address such variety of issues, many new methods, applications, and standards are currently being developed in the area of big data analytics at a rapid pace. Some of these have been covered in the chapters of the present title.

2 Computational versus Statistical Complexity

Interestingly, the computer scientists and the statisticians—the two communities of researchers that are perhaps most directly affected by the phenomenon of big data— have, for cultural reasons, adopted distinct initial stances in response to it. The primary concern of the computer scientist—who must design efficient data and file structures to store massive datasets and implement algorithms on them—stems from computational complexity. It concerns the required number of computing steps to solve a given problem whose complexity is defined in terms of the length of input data as represented by a reasonable encoding mechanism (say, a N bit binary string).

Therefore, as data volume increases, any method that requires significantly more than $O(N \log(N))$ steps (i.e., exceeding the order of time that a single pass over the full data would require) could be impractical. While some of the important problems in practice with $O(N \log(N))$ solutions are just about scalable (e.g., Fast Fourier transform), those of higher complexity, certainly including the NP-Complete class of problems, would require help from algorithmic strategies like approximation, randomization, sampling, etc. Thus, while classical complexity theory may consider polynomial time solutions as the hallmark of computational tractability, the world of big data is indeed even more demanding.

Big data are being collected in a great variety of ways, types, shapes, and sizes. The data dimensionality p and the number of data points or sample size n are usually the main components in characterization of data volume. Interestingly, big p small n datasets may require a somewhat different set of analytical tools as compared to big n big p data. Indeed, there may not be a single method that performs well on all types of big data. Five aspects of the data matrix are important [2]:

(i) the dimension p representing the number of explanatory variables measured;
(ii) the sample size n representing the number of observations at which the variables are measured or collected;
(iii) the relationship between n and p measured by their ratio;
(iv) the type of variables measured (categorical, interval, count, ordinal, real-valued, vector-valued, function-valued) and the indication of scales or units of measurement; and
(v) the relationship among the columns of the data matrix to check multicollinearity in the explanatory variables.

To characterize big data analytics as different from (or extension of) usual data analysis, one could suggest various criteria, especially if the existing analytical strategies are not adequate for the solving the problem in hand due to certain properties of data. Such properties could go beyond sheer data volume. High data velocity can present unprecedented challenges to a statistician who may not be used to the idea of forgoing (rather than retaining) data points, as they stream out, in order to satisfy computational constraints such as single pass (time constraint) and bounded storage (space constraint). High data variety may require multidisciplinary insights to enable one to make sensible inference based on integration of seemingly unrelated datasets. On one hand, such issues could be viewed merely as cultural gaps, while on the other, they can motivate the development of the necessary formalisms that can bridge those gaps. Thereby, a better understanding of the pros and cons of different algorithmic choices can help an analyst decide about the most suitable of the possible solution(s) objectively. For instance, given a p variable dataset, a time-data complexity class can be defined in terms of $n(p)$, $r(p)$ and $t(p)$ to compare the performance tradeoffs among the different choices of algorithms to solve a particular big data problem within a certain number of samples $n(p)$, a certain level of error or risk $r(p)$ and a certain amount of time $t(p)$ [3].

While a computer scientist may view data as physical entity (say a string having physical properties like length), a statistician is used to viewing data points as instances of an underlying random process for data generation, typically modeled using suitable probability distributions. Therefore, by assuming such underlying structure, one could view the growing number of data points as a potential source of simplification of that structural complexity. Thus, bigger n can lead to, in a classical statistical framework, favorable conditions under which inference based on the assumed model can be more accurate, and model asymptotics can possibly hold [3]. Similarly, big p may not always be viewed unfavorably by the statistician, say, if the model-fitting task can take advantage of data properties such as variable sparsity whereby the coefficients—say, of a linear model—corresponding to many variables, except for perhaps a few important predictors, may be shrunk towards zero [4]. In particular, it is the big p and small n scenario that can challenge key assumptions made in certain statistical tests of hypotheses. However, while data analytics shifts from a static hypothesis-driven approach to a more exploratory or dynamic large data-driven one, the computational concerns, such as how to decide each step of an analytical pipeline, of both the computer scientist and the statistician have gradually begun to converge.

Let us suppose that we are dealing with a multiple linear regression problem with p explanatory variables under Gaussian error. For a model space search for variable selection, we have to find the best subset from among $2^p - 1$ sub-models. If $p = 20$, then $2^p - 1$ is about a million; but if $p = 40$, then the same increases to about a trillion! Hence, any problem with more than $p = 50$ variables is potentially a big data problem. With respect to n, on the other hand, say, for linear regression methods, it takes $O(n^3)$ number of operations to invert an $n \times n$ matrix. Thus, we might say that a dataset is big n if $n > 1000$. Interestingly, for a big dataset, the ratio n/p could be even more important than the values of n and p taken separately. According to a recent categorization [2], it is information-abundant if $n/p \geq 10$, information-scarce if $1 \leq n/p < 10$, and information-poor if $n/p < 1$.

Theoretical results, e.g., [5], show that the data dimensionality p is not considered as "big" relative to n unless p dominates $\sqrt{n}$ asymptotically. If $p \gg n$, then there exists a multiplicity of solutions for an optimization problem involving model-fitting, which makes it ill-posed. Regularization methods such as the Lasso (cf. Tibshirani [6]) are used to find a feasible optimal solution, such that the regularization term offers a tradeoff between the error of the fit model and its complexity. This brings us to the non-trivial issues of model tuning and evaluation when it comes to big data. A "model-complexity ladder" might be useful to provide the analyst with insights into a range of possible models to choose from, often driven by computational considerations [7]. For instance, for high-dimensional data classification, the modeling strategies could range from, say, naïve Bayes and logistic regression and moving up to, possibly, hierarchical nonparametric Bayesian approaches. Ideally, the decision to select a complex model for a big dataset should be a careful one that is justified by the signal-to-noise ratio of the dataset under consideration [7].

A more complex model may overfit the training data, and thus predict poorly for test data. While there has been extensive research on how the choice of a model with unnecessarily high complexity could be penalized, such tradeoffs are not quite well understood for different types of big data, say, involving streams with nonstationary characteristics. If the underlying data generation process changes, then the data complexity can change dynamically. New classes can emerge or disappear from data (also known as "concept drift") even while the model-fitting is in progress. In a scenario where the data complexity can change, one might opt for a suitable nonparametric model whose complexity and number of parameters could also grow as more data points become available [7]. For validating a selected model, cross-validation is still very useful for high-dimensional data. For big data, however, a single selected model does not typically lead to optimal prediction. If there is multicollinearity among the variables, which is possible when p is big, the estimators will be unstable and have large variance. Bootstrap aggregation (or bagging), based on many resamplings of size n, can reduce the variance of the estimators by aggregation of bootstrapped versions of the base estimators. For big n, the "bag of small bootstraps" approach can achieve similar effects by using smaller subsamples of the data. It is through such useful adaptations of known methods for "small" data that a toolkit based on fundamental algorithmic strategies has now evolved and is being commonly applied to big data analytics, and we mention some of these below.

3   Techniques to Cope with Big Data

Sampling, the general process of selecting a subset of data points from a given input, is among the most established and classical techniques in statistics, and proving to be extremely useful in making big data tractable for analytics. Random sampling strategies are commonly used in their simple, stratified, and numerous other variants for their effective handling of big data. For instance, the classical Fisher–Yates shuffling is used for reservoir sampling in online algorithms to ensure that for a given "reservoir" sample of k points drawn from a data stream of big but unknown size n, the probability of any new point being included in the sample remains fixed at k/n, irrespective of the value of the new data. Alternatively, there are case-based or event-based sampling approaches for detecting special cases or events of interest in big data. Priority sampling is used for different applications of stream data. The very fast decision tree (VFDT) algorithm allows big data classification based on a tree model that is built like CART but uses subsampled data points to make its decisions at each node of the tree. A probability bound (e.g., the Hoeffding inequality) ensures that had the tree been built instead using the full dataset, it would not differ by much from the model that is based on sampled data [8]. That is, the sequence of decisions (or "splits") taken by both trees would be similar on a given dataset with probabilistic performance guarantees. Given the fact that big data (say, the records on the customers of a particular brand) are not necessarily generated by random sampling of the population, one must be careful about possible selection bias in the identification of various classes that are present in the population.

Massive amounts of data are accumulating in social networks such as Google, Facebook, Twitter, LinkedIn, etc. With the emergence of big graph data from social networks, astrophysics, biological networks (e.g., protein interactome, brain connectome), complex graphical models, etc., new methods are being developed for sampling large graphs to estimate a given network's parameters, as well as the node-, edge-, or subgraph-statistics of interest. For example, snowball sampling is a common method that starts with an initial sample of seed nodes, and in each step i, it includes in the sample all nodes that have edges with the nodes in the sample at step $i - 1$ but were not yet present in the sample. Network sampling also includes degree-based or PageRank-based methods and different types of random walks. Finally, the statistics are aggregated from the sampled subnetworks. For dynamic data streams, the task of statistical summarization is even more challenging as the learnt models need to be continuously updated. One approach is to use a "sketch", which is not a sample of the data but rather its synopsis captured in a space-efficient representation, obtained usually via hashing, to allow rapid computation of statistics therein such that a probability bound may ensure that a high error of approximation is unlikely. For instance, a sublinear count-min sketch could be used to determine the most frequent items in a data stream [9]. Histograms and wavelets are also used for statistically summarizing data in motion [8].

The most popular approach for summarizing large datasets is, of course, clustering. Clustering is the general process of grouping similar data points in an unsupervised manner such that an overall aggregate of distances between pairs of points within each cluster is minimized while that across different clusters is maximized. Thus, the cluster-representatives, (say, the k means from the classical k-means clustering solution), along with other cluster statistics, can offer a simpler and cleaner view of the structure of the dataset containing a much larger number of points ($n \gg k$) and including noise. For big n, however, various strategies are being used to improve upon the classical clustering approaches. Limitations, such as the need for iterative computations involving a prohibitively large $O(n^2)$ pairwise-distance matrix, or indeed the need to have the full dataset available beforehand for conducting such computations, are overcome by many of these strategies. For instance, a two-step online–offline approach (cf. Aggarwal [10]) first lets an online step to rapidly assign stream data points to the closest of the k' ($\ll n$) "microclusters." Stored in efficient data structures, the microcluster statistics can be updated in real time as soon as the data points arrive, after which those points are not retained. In a slower offline step that is conducted less frequently, the retained k' microclusters' statistics are then aggregated to yield the latest result on the $k(< k')$ actual clusters in data. Clustering algorithms can also use sampling (e.g., CURE [11]), parallel computing (e.g., PKMeans [12] using MapReduce) and other strategies as required to handle big data.

While subsampling and clustering are approaches to deal with the big n problem of big data, dimensionality reduction techniques are used to mitigate the challenges of big p. Dimensionality reduction is one of the classical concerns that can be traced back to the work of Karl Pearson, who, in 1901, introduced principal component analysis (PCA) that uses a small number of principal components to explain much of the variation in a high-dimensional dataset. PCA, which is a lossy linear model of dimensionality reduction, and other more involved projection models, typically using matrix decomposition for feature selection, have long been the mainstays of high-dimensional data analysis. Notably, even such established methods can face computational challenges from big data. For instance, $O(n^3)$ time-complexity of matrix inversion, or implementing PCA, for a large dataset could be prohibitive in spite of being polynomial time—i.e. so-called "tractable"—solutions.

Linear and nonlinear multidimensional scaling (MDS) techniques–working with the matrix of $O(n^2)$ pairwise-distances between all data points in a high-dimensional space to produce a low-dimensional dataset that preserves the neighborhood structure—also face a similar computational challenge from big n data. New spectral MDS techniques improve upon the efficiency of aggregating a global neighborhood structure by focusing on the more interesting local neighborhoods only, e.g., [13]. Another locality-preserving approach involves random projections, and is based on the Johnson–Lindenstrauss lemma [14], which ensures that data points of sufficiently high dimensionality can be "embedded" into a suitable low-dimensional space such that the original relationships between the points are approximately preserved. In fact, it has been observed that random projections may make the distribution of points more Gaussian-like, which can aid clustering of the projected points by fitting a finite mixture of Gaussians [15]. Given the randomized nature of such embedding, multiple projections of an input dataset may be clustered separately in this approach, followed by an ensemble method to combine and produce the final output.

The term "curse of dimensionality" (COD), originally introduced by R.E. Bellman in 1957, is now understood from multiple perspectives. From a geometric perspective, as p increases, the exponential increase in the volume of a p-dimensional neighborhood of an arbitrary data point can make it increasingly sparse. This, in turn, can make it difficult to detect local patterns in high-dimensional data. For instance, a nearest-neighbor query may lose its significance unless it happens to be limited to a tightly clustered set of points. Moreover, as p increases, a "deterioration of expressiveness" of the L_p norms, especially beyond L_1 and L_2, has been observed [16]. A related challenge due to COD is how a data model can distinguish the few relevant predictor variables from the many that are not, i.e., under the condition of dimension sparsity. If all variables are not equally important, then using a weighted norm that assigns more weight to the more important predictors may mitigate the sparsity issue in high-dimensional data and thus, in fact, make COD less relevant [4].

In practice, the most trusted workhorses of big data analytics have been parallel and distributed computing. They have served as the driving forces for the design of most big data algorithms, softwares and systems architectures that are in use today. On the systems side, there is a variety of popular platforms including clusters, clouds, multicores, and increasingly, graphics processing units (GPUs). Parallel and distributed databases, NoSQL databases for non-relational data such as graphs and documents, data stream management systems, etc., are also being used in various applications. BDAS, the Berkeley Data Analytics Stack, is a popular open source software stack that integrates software components built by the Berkeley AMP Lab (and third parties) to handle big data [17]. Currently, at the base of the stack, it starts with resource virtualization by Apache Mesos and Hadoop Yarn, and uses storage systems such as Hadoop Distributed File System (HDFS), Auxilio (formerly Tachyon) and Succinct upon which the Spark Core processing engine provides access and interfaces to tasks like data cleaning, stream processing, machine learning, graph computation, etc., for running different applications, e.g., cancer genomic analysis, at the top of the stack.

Some experts anticipate a gradual convergence of architectures that are designed for big data and high-performance computing. Important applications such as large simulations in population dynamics or computational epidemiology could be built on top of these designs, e.g., [18]. On the data side, issues of quality control, standardization along with provenance and metadata annotation are being addressed. On the computing side, various new benchmarks are being designed and applied. On the algorithmic side, interesting machine learning paradigms such as deep learning and advances in reinforcement learning are gaining prominence [19]. Fields such as computational learning theory and differential privacy will also benefit big data with their statistical foundations. On the applied statistical side, analysts working with big data have responded to the need of overcoming computational bottlenecks, including the demands on accuracy and time. For instance, to manage space and achieve speedup when modeling large datasets, a "chunking and averaging" strategy has been developed for parallel computation of fairly general statistical estimators [4]. By partitioning a large dataset consisting of n i.i.d. samples (into r chunks each of manageable size $\lfloor n/r \rfloor$), and computing the estimator for each individual chunk of data in a parallel process, it can be shown that the average of these chunk-specific estimators has comparable statistical accuracy as the estimate on the full dataset [4]. Indeed, superlinear speedup was observed in such parallel estimation, which, as n grows larger, should benefit further from asymptotic properties.

4 Conclusion

In the future, it is not difficult to see that perhaps under pressure from the myriad challenges of big data, both the communities—of computer scientists and statistics—may come to share mutual appreciation of the risks, benefits and tradeoffs faced by each, perhaps to form a new species of data scientists who will be better equipped with dual forms of expertise. Such a prospect raises our hopes to address some of the "giant" challenges that were identified by the National Research Council of the National Academies in the United States in its 2013 report titled '*Frontiers in Massive Data Analysis*'. These are (1) basic statistics, (2) generalized N-body problem, (3) graph-theoretic computations, (4) linear algebraic computations, (5) optimization, (6) integration, and (7) alignment problems. (The reader is encouraged to read this insightful report [7] for further details.) The above list, along with the other present and future challenges that may not be included, will continue to serve as a reminder that a long and exciting journey lies ahead for the researchers and practitioners in this emerging field.

References

1. Kennedy R, King G, Lazer D, Vespignani A (2014) The parable of google flu. Traps in big data analysis. Science 343:1203–1205
2. Fokoue E (2015) A taxonomy of Big Data for optimal predictive machine learning and data mining. arXiv.1501.0060v1 [stat.ML] 3 Jan 2015
3. Chandrasekaran V, Jodan MI (2013) Computational and statistical tradeoffs via convex relaxation. Proc Natl Acad Sci USA 110:E1181–E1190
4. Matloff N (2016) Big n versus big p in Big data. In: Bühlmann P, Drineas P (eds) Handbook of Big Data. CRC Press, Boca Raton, pp 21–32
5. Portnoy S (1988) Asymptotic behavior of likelihood methods for exponential families when the number of parameters tends to infinity. Ann Stat 16:356–366
6. Tibshirani R (1996) Regression analysis and selection via the lasso. J R Stat Soc Ser B 58:267–288
7. Report of National Research Council (2013) Frontiers in massive data analysis. National Academies Press, Washington D.C
8. Gama J (2010) Knowledge discovery from data streams. Chapman Hall/CRC, Boca Raton
9. Cormode G, Muthukrishnan S (2005) An improved data stream summary: the count-min sketch and its applications. J Algorithms 55:58–75
10. Aggarwal C (2007) Data streams: models and algorithms. Springer, Berlin
11. Rastogi R, Guha S, Shim K (1998) Cure: an efficient clustering algorithm for large databases. In: Proceedings of the ACM SIGMOD, pp 73–84
12. Ma H, Zhao W, He C (2009) Parallel k-means clustering based on MapReduce. CloudCom, pp 674–679
13. Aflalo Y, Kimmel R (2013) Spectral multidimensional scaling. Proc Natl Acad Sci USA 110:18052–18057
14. Johnson WB, Lindenstrauss J (1984) Extensions of lipschitz mappings into a hilbert space. Contemp Math 26:189–206
15. Fern XZ, Brodley CE (2003) Random projection for high dimensional data clustering: a cluster ensemble approach. In: Proceedings of the ICML, pp 186–193
16. Zimek A (2015) Clustering high-dimensional data. In: Data clustering: algorithms and applications. CRC Press, Boca Raton
17. University of California at Berkeley AMP Lab. https://amplab.cs.berkeley.edu/. Accessed April 2016
18. Pyne S, Vullikanti A, Marathe M (2015) Big data applications in health sciences and epidemiology. In: Raghavan VV, Govindaraju V, Rao CR (eds) Handbook of statistics, vol 33. Big Data analytics. Elsevier, Oxford, pp 171–202
19. Jordan MI, Mitchell TM (2015) Machine learning: trends, perspectives and prospects. Science 349(255–60):26

Massive Data Analysis: Tasks, Tools, Applications, and Challenges

Murali K. Pusala, Mohsen Amini Salehi, Jayasimha R. Katukuri, Ying Xie
and Vijay Raghavan

Abstract In this study, we provide an overview of the state-of-the-art technologies
in programming, computing, and storage of the massive data analytics landscape.
We shed light on different types of analytics that can be performed on massive data.
For that, we first provide a detailed taxonomy on different analytic types along with
examples of each type. Next, we highlight technology trends of massive data analytics that are available for corporations, government agencies, and researchers. In
addition, we enumerate several instances of opportunities that exist for turning massive data into knowledge. We describe and position two distinct case studies of massive data analytics that are being investigated in our research group: recommendation
systems in e-commerce applications; and link discovery to predict unknown association of medical concepts. Finally, we discuss the lessons we have learnt and open
challenges faced by researchers and businesses in the field of massive data analytics.

M.K. Pusala · J.R. Katukuri · V. Raghavan (✉)
Center of Advanced Computer Studies (CACS), University of Louisiana Lafayette,
Lafayette, LA 70503, USA
e-mail: vijay@cacs.louisiana.edu

M.K. Pusala
e-mail: mxp6168@cacs.louisiana.edu

J.R. Katukuri
e-mail: jaykatukuri@gmail.com

M. Amini Salehi
School of Computing and Informatics, University of Louisiana Lafayette,
Lafayette, LA 70503, USA
e-mail: amini@louisiana.edu

Y. Xie
Department of Computer Science, Kennesaw State University,
Kennesaw, GA 30144, USA
e-mail: yxie2@kennesaw.edu

© Springer India 2016
S. Pyne et al. (eds.), *Big Data Analytics*, DOI 10.1007/978-81-322-3628-3_2

1 Introduction

1.1 Motivation

Growth of Internet usage in the last decade has been at an unprecedented rate from 16 million, which is about 0.4 % of total population in 1995, to more than 3 billion users, which is about half of the world's population in mid-2014. This revolutionized the way people communicate and share their information. According to [46], just during 2013, 4.4 zettabytes (4.4×2^{70} bytes) of information have created and replicated, and it estimated to grow up to 44 zettabytes by 2020. Below, we explain few sources from such massive data generation.

Facebook[1] has an average of 1.39 billion monthly active users exchanging billions of messages and postings every day [16]. There is also a huge surge in multimedia content like photos and videos. For example, in popular photo sharing social network Instagram,[2] on average, 70 million photos uploaded and shared every day [27]. According to other statistics published by Google on its video streaming service, YouTube,[3] has approximately 300 h of video uploaded every minute and billions of views generated every day [62].

Along with Individuals, organizations are also generating a huge amount of data, mainly due to increased use of networked sensors in various sectors of organizations. For example, by simply replacing traditional bar code systems with radio frequency identification (RFID) systems organizations have generated 100 to 1000 times more data [57].

Organization's interest on customer behavior is another driver for producing massive data. For instance, Wal-Mart[4] handles more than a million customer transactions each hour and maintains a database that holds more than 2.5 petabytes of data [57]. Many businesses are creating a 360° view of a customer by combining transaction data with social networks and other sources.

Data explosion is not limited to individuals or organizations. With the increase of scientific equipment sensitivity and advancements in technology, the scientific and research, community is also generating a massive amount of data. Australian Square Kilometer Array Pathfinder radio telescope [8] has 36 antennas streams approximately 250 GB of data per second per antenna that collectively produces nine terabytes of data per second. In another example, particle accelerator, particle detector, and simulations at Large Hadron Collider (LHC) at CERN [55] generate approximately 15 petabytes of data per year.

---

[1]https://facebook.com.
[2]https://instagram.com.
[3]http://www.youtube.com.
[4]http://www.walmart.com.

1.2 Big Data Overview

The rapid explosion of data is usually referred as *"Big Data"*, which is a trending topic in both industry and academia. Big data (aka Massive Data) is defined as, data that cannot be handled or analyzed by conventional processing and storage tools. Big data is also characterized by features, known as *5V's*. These features are: *volume, variety, velocity, variability,* and *veracity* [7, 21].

Traditionally, most of the available data is structured data and stored in conventional databases and data warehouses for supporting all kinds of data analytics. With the Big data, data is no longer necessarily structured. Instead, it contains a *variety* of data sources, including structured, semi-structured, and unstructured data [7]. It is estimated that 85 % of total organizational data are unstructured data [57] and almost all the data generated by individuals (e.g., emails, messages, blogs, and multimedia) are unstructured data too. Traditional relational databases are no longer a viable option to store text, video, audio, images, and other forms of unstructured data. This creates a need for special types of NoSQL databases and advanced analytic methods.

Velocity of data is described as problem of handling and processing data at the speeds at which they are generated to extract a meaningful value. Online retailers store every attribute (e.g., clicks, page visits, duration of visits to a page) of their customers' visits to their online websites. There is a need to analyze customers' visits within a reasonable timespan (e.g., real time) to recommend similar items and related items with respect to the item a customer is looking at. This helps companies to attract new customers and keep an edge over their competitors. Some organizations analyze data as a stream in order to reduce data storage. For instance, LHC at CERN [55] analyzes data before storing to meet the storage requirements. Smart phones are equipped with modern location detection sensors that enable us to understand the customer behavior while, at the same time, creating the need for real-time analysis to deliver location-based suggestions.

Data *variability* is the variation in data flow with time of day, season, events, etc. For example, retailers sell significantly more in November and December compared to rest of year. According to [1], traffic to retail websites surges during this period. The challenge, in this scenario, is to provide resources to handle sudden increases in users' demands. Traditionally, organizations were building in-house infrastructure to support their peak-estimated demand periods. However, it turns out to be costly, as the resources will remain idle during the rest of the time. However, the emergence of advanced distributed computing platforms, known as 'the cloud,' can be leveraged to enable on-demand resource provisioning through third party companies. Cloud provides efficient computational, storage, and other services to organizations and relieves them from the burden of over-provisioning resources [49].

Big data provides advantage in decision-making and analytics. However, among all data generated in 2013 only 22 % of data are tagged, or somehow characterized as useful data for analysis, and only 5 % of data are considered valuable or "Target Rich" data. The quality of collected data, to extract a value from, is referred as *veracity*. The ultimate goal of an organization in processing and analyzing data is to obtain hidden information in data. Higher quality data increases the likelihood of effective decision-making and analytics. A McKinsey study found that retailers using full potential from Big data could increase the operating margin up to 60 % [38]. To reach this goal, the quality of collected data needs to be improved.

1.3 Big Data Adoption

Organizations have already started tapping into the potential of Big data. Conventional data analytics are based on structured data, such as the transactional data, that are collected in a data warehouse. Advanced massive data analysis helps to combine traditional data with data from different sources for decision-making. Big data provides opportunities for analyzing customer behavior patterns based on customer actions inside (e.g., organization website) and outside (e.g., social networks).

In a manufacturing industry, data from sensors that monitor machines' operation are analyzed to predict failures of parts and replace them in advance to avoid significant down time [25]. Large financial institutions are using Big data analytics to identify anomaly in purchases and stop frauds or scams [3].

In spite of the wide range of emerging applications for Big data, organizations are still facing challenges to adopt Big data analytics. A report from AIIM [9], identified three top challenges in the adoption of Big data, which are lack of skilled workers, difficulty to combine structured and unstructured data, and security and privacy concerns. There is a sharp rise in the number of organizations showing interest to invest in Big data related projects. According to [18], in 2014, 47 % of organizations are reportedly investing in Big data products, as compared to 38 % in 2013. IDC predicted that the Big data service market has reached 11 billion dollars in 2009 [59] and it could grow up to 32.4 billion dollars by end of 2017 [43]. Venture capital funding for Big data projects also increased from 155 million dollars in 2009 to more than 893 million dollars in 2013 [59].

1.4 The Chapter Structure

From the late 1990s, when Big data phenomenon was first identified, until today, there has been many improvements in computational capabilities, storage devices have become more inexpensive, thus, the adoption of data-centric analytics has increased. In this study, we provide an overview of Big data analytic types, offer insight into Big data technologies available, and identify open challenges.

The rest of this paper is organized as following. In Sect. 2, we explain different categories of Big data analytics, along with application scenarios. Section 3 of the chapter describes Big data computing platforms available today. In Sect. 4, we provide some insight into the storage of huge volume and variety data. In that section, we also discuss some commercially available cloud-based storage services. In Sect. 5, we present two real-world Big data analytic projects. Section 6 discusses open challenges in Big data analytics. Finally, we summarize and conclude the main contributions of the chapter in Sect. 7.

2 Big Data Analytics

Big data analytics is the process of exploring Big data, to extract hidden and valuable information and patterns [48]. Big data analytics helps organizations in more informed decision-making. Big data analytics applications can be broadly classified as *descriptive*, *predictive*, and *prescriptive*. Figure 1 illustrates the data analytic classes, techniques, and example applications. In the rest of this section, with reference to Fig. 1, we elaborate on these Big data analytic types.

2.1 Descriptive Analytics

Descriptive analytics mines massive data repositories to extract potential patterns existing in the data. Descriptive analytics drills down into historical data to detect patterns like variations in operating costs, sales of different products, customer buying preferences, etc.

Typically it is the first step of analytics in decision-making, answering the question of "what has happened? ". It summarizes raw data into a human understandable format. Most of the statistical analysis used in day-to-day Business Intelligence (BI) regarding a company's production, financial operations, sales, inventory, and customers come under descriptive analytics [61]. Analytics involve simple techniques, such as regression to find correlation among various variables and drawing charts,

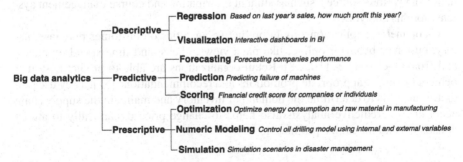

Fig. 1 Types of Big data analytics: The second level in the hierarchy is the categorization of analytics. The third level, explains the typical techniques, and provides example in the corresponding analytic category

to identify trends in the data, and visualize data in a meaningful and understandable way, respectively.

For example, Dow Chemicals used descriptive analytics to identify under-utilized space in its offices and labs. As a result, they were able to increase space utilization by 20 % and save approximately $4 million annually [14].

2.2 Predictive Analytics

With descriptive analytics, organizations can understand what happened in the past. However, at a higher level of decision-making is to address the question of "what could happen?". Predictive analytics helps to combine massive data from different sources with the goal of predicting future trends or events. Predictive analytics evaluates the future, by forecasting trends, by generating prediction models, and by scoring.

For example, industries use predictive analytics to predict machine failures using streaming sensor data [25]. Organizations are able to forecast their sales trends or overall performance [35]. Financial institutions devote a lot of resources to predict credit risk scores for companies or individuals. Eventhough predictive analytics cannot predict with 100 % certainty, but it helps the companies in estimating future trends for more informed decision-making.

Southwest airlines has partnered with National Aeronautics and Space Administration (NASA) to work on a Big data-mining project [42]. They apply text-based analysis on data from sensors in their planes in order to find patterns that indicate potential malfunction or safety issues.

Purdue University uses Big data analytics to predict academic and behavioral issues [45]. For each student, the system predicts and generates a risk profile indicating how far a student succeeds in a course and labels the risk levels as green (high probability of success), yellow (potential problems), and red (risk of failure) by using data from various sources, such as student information and course management systems for this analytics.

E-commerce applications apply predictive analytics on customer purchase history, customer behavior online, like page views, clicks, and time spend on pages, and from other sources [10, 58]. Retail organizations are able to predict customer behavior to target appropriate promotions and recommendations [31]. They use predictive analysis to determine the demand of inventory and maintain the supply chain accordingly. Predictive analysis also helps to change price dynamically to attract consumers and maximize profits [2].

2.3 Prescriptive Analytics

Descriptive and predictive analytics helps to understand the past and predict the future. The next stage in decision-making is "how can we make it happen?"—the answer is prescriptive analytics. The goal of prescriptive analytics is to assist professionals in assessing the impact of different possible decisions. It is a relatively new analytic method. According to Gartner [19], only 3 % of companies use prescriptive analytics in their decision-making. Prescriptive analytics involves techniques such as optimization, numerical modeling, and simulation.

Oil and Gas exploration industries use prescriptive analytics to *optimize* the exploration process. Explorers are using massive datasets from different sources in the exploration process and use prescriptive analytics to optimize drilling location [56]. They use earth's sedimentation characteristics, temperature, pressure, soil type, depth, chemical composition, molecular structures, seismic activity, machine data, and others to determine the best possible location to drill [15, 17]. This helps to optimize selection of drilling location, and avoid the cost and effort of unsuccessful drills.

Health care is one of the sectors benefiting from applying Big data prescriptive analytics. Prescriptive analytics can recommend diagnoses and treatments to a doctor by analyzing patient's medical history, similar conditioned patient's history, allergies, medicines, environmental conditions, stage of cure, etc. According to [54], the Aurora Health Care Center saves six million USD annually by using Big data analytics and recommending best possible treatment to doctors.

3 Big Data Analytics Platforms

There are several Big data analytics platforms available. In this section, we present advances within the Big data analytics platforms.

3.1 MapReduce

MapReduce framework represents a pioneering schema for performing Big data analytics. It has been developed for a dedicated platform (such as a cluster). MapReduce framework has been implemented in three different ways. The first implementation was achieved by Google [13] under a proprietary license. The other two implementations are: Hadoop [33] and Spark [66],which are available as open source. There are other platforms that, in fact, stem from these basic platforms.

The core idea of MapReduce is based on developing two input functions namely, Map and Reduce. Programmers need to implement these functions. Each of these functions utilizes the available resources to process Big data in parallel. The MapReduce works closely with a distributed storage system to carry out operations such as storing input, intermediate, and output data. Distributed file systems, such as Hadoop Distributed File System (HDFS) [52] and Google File System (GFS), have been developed to the MapReduce framework [20].

Every MapReduce workflow typically contains three steps (phases) namely, Mapping step, Shuffling step, and Reduce step. In the Map step, user (programmer) implements the functionality required in the Map function. The defined Map function will be executed against the input dataset across the available computational resources. The original (i.e., input) data are partitioned and placed in a distributed file system (DFS). Then, each Map task processes a partition of data from the DFS and generates intermediate data that are stored locally on the worker machines where the processing was taking place.

Distributing the intermediate data on the available computational resources is required to enable parallel Reduce. This step is known as Shuffling. The distribution of the intermediate data is performed in an all-to-all fashion that generally creates a communication bottleneck. Once the distribution of intermediate data is performed, the Reduce function is executed to produce the output, which is the final result of the MapReduce processing. Commonly, developers create a chain of MapReduce jobs (also referred to as a multistage MapReduce job), such as the Yahoo! WebMap [5]. In this case, the output of one MapReduce job is consumed as the intermediate data for the next MapReduce job in the chain.

3.2   Apache Hadoop

Hadoop [33] framework was developed as an open source product by Yahoo! and widely adopted for Big data analytics by the academic and industrial communities. The main design advantage of Hadoop is its fault-tolerance. In fact, Hadoop has been designed with the assumption of failure as a common issue in distributed systems. Therefore, it is robust against failures commonly occur during different phases of execution.

Hadoop Distributed File System (HDFS) and MapReduce are two main building blocks of Hadoop. The former is the storage core of Hadoop (see Sect. 4.1 for details). The latter, MapReduce engine, is above the file system and takes care of executing the application by moving binaries to the machines that have the related data.

For the sake of fault-tolerance, HDFS replicates data blocks in different racks; thus, in case of failure in one rack, the whole process would not fail. A Hadoop cluster includes one master node and one or more worker nodes. The master node includes four components namely, JobTracker, TaskTracker, NameNode, and DataNode. The worker node just includes DataNode and TaskTracker. The JobTracker receives user applications and allocates them to available TaskTracker nodes, while considering data locality. JobTracker assures about the health of TaskTrackers based on regular heartbeats it receives from them. Although Hadoop is robust against failures in a distributed system, its performance is not the best amongst other available tools because of frequent disk accesses [51].

3.3 Spark

Spark is a more recent framework developed at UC Berkeley [66]. It is being used for research and production applications. Spark offers a general-purpose programming interface in the Scala programming language for interactive, in-memory data analytics of large datasets on a cluster.

Spark provides three data abstractions for programming clusters namely, *resilient distributed datasets (RDDs)*, *broadcast variables*, and *accumulators*. RDD is a read-only collection of objects partitioned across a set of machines. It can reconstruct lost partitions or recover in the event of a node failure. RDD uses a restricted shared memory to achieve fault-tolerance. Broadcast variables and accumulators are two restricted types of shared variables. Broadcast variable is a shared object wrapped around a read-only value, which ensures it is only copied to each worker once. Accumulators are shared variables with an *add* operation. Only workers can perform an operation on an accumulator and only users' driver programs can read from it. Eventhough, these abstractions are simple and limited, they can be used to develop several cluster-based applications.

Spark uses master/slave architecture. It has one master instance, which runs a user-defined driver program. At run-time, the driver program launches multiple workers in the cluster, which read data from the shared filesystem (e.g., Hadoop Distributed File System). Workers create RDDs and write partitions on RAM as defined by the driver program. Spark supports RDD transformations (e.g., map, filter) and actions (e.g., count, reduce). Transformations generate new datasets and actions return a value, from the existing dataset.

Spark has proved to be 20X faster than Hadoop for iterative applications, was shown to speed up a real-world data analytics report by 40X, and has been used interactively to scan a 1 TB dataset with 57 s latency [65].

3.4 High Performance Computing Cluster

LexisNexis Risk Solutions originally developed High Performance Computing Cluster (HPCC),[5] as a proprietary platform, for processing and analyzing large volumes of data on clusters of commodity servers more than a decade ago. It was turned into an open source system in 2011. Major components of an HPCC system include a Thor cluster and a Roxie cluster, although the latter is optional. Thor is called the data refinery cluster, which is responsible for extracting, transforming, and loading

---

[5]http://hpccsystems.com.

(ETL), as well as linking and indexing massive data from different sources. Roxie is called the query cluster, which is responsible for delivering data for online queries and online analytical processing (OLAP).

Similar to Hadoop, HPCC also uses a distributed file system to support parallel processing on Big data. However, compared with HDFS, the distributed file system used by HPCC has some significant distinctions. First of all, HPCC uses two types of distributed file systems; one is called Thor DFS that is intended to support Big data ETL in the Thor cluster; the other is called Roxie DFS that is intended to support Big data online queries in the Roxie cluster. Unlike HDFS that is key-value pair based, the Thor DFS is record-oriented, which is flexible enough to support data sets of different formats, such as CSV, XML, fixed or variable length of records, and records with nested structures. Thor DFS distributes a file across all nodes in the Thor cluster with an even number of records for each node. The Roxie DFS uses distributed B+ tree for data indexing to support efficient delivery of data for user queries.

HPCC uses a data-centric, declarative programming language called Enterprise Control Language (ECL) for both data refinery and query delivery. By using ECL, the user specifies what needs to be done on data instead of how to do it. The data transformation in ECL can be specified either locally or globally. Local transformation is carried out on each file part stored in a node of the Thor cluster in a parallel manner, whereas global transformation processes the global data file across all nodes of the Thor cluster. Therefore, HPCC not only pioneers the current Big data computing paradigm that moves computing to where the data is, but also maintains the capability of processing data in a global scope. ECL programs can be extended with C++ libraries and compiled into optimized C++ code. A performance comparison of HPCC with Hadoop shows that, on a test cluster with 400 processing nodes, HPCC is 3.95 faster than Hadoop on the Terabyte Sort benchmark test [41]. One of the authors of this chapter is currently conducting a more extensive performance comparison of HPCC and Hadoop on a variety of Big data analysis algorithms. More technical details on HPCC can be found in [24, 40, 41, 47].

4 Distributed Data Management Systems for Big Data Analytics

As we discussed earlier in this chapter, huge volumes and a variety of data create a need for special types of data storage. In this section, we discuss recent advances in storage systems for Big data analytics and some commercially available cloud-based storage services.

4.1 Hadoop Distributed File System

The Hadoop Distributed File System (HDFS)[6] is a distributed file system designed to run reliably and to scale on commodity hardware. HDFS achieves high fault-tolerance by dividing data into smaller chunks and replicating them across several nodes in a cluster. It can scale up to 200 PB in data, and 4500 machines in single cluster. HDFS is a side project of Hadoop and works closely with it.

HDFS is designed to work efficiently in batch mode, rather than in interactive mode. Characteristics of typical applications developed for HDFS, such as write once and read multiple times, and simple and coherent data access, increases the throughput. HDFS is designed to handle large file sizes from Gigabytes to a few Terabytes.

HDFS follows the master/slave architecture with one NameNode and multiple DataNodes. NameNode is responsible for managing the file system's meta data and handling requests from applications. DataNodes physically hold the data. Typically, every node in the cluster has one DataNode. Every file stored in HDFS is divided into blocks with default block size of 64 MB. For the sake of fault tolerance, every block is replicated into user-defined number of times (recommended to be a minimum of 3 times) and distributed across different data nodes. All meta data about replication and distribution of the file are stored in the NameNode. Each DataNode sends a heartbeat signal to NameNode. If it fails to do so, the NameNode marks the DataNode as failed.

HDFS maintains a Secondary NameNode, which is periodically updated with information from NameNode. In case of NameNode failure, HDFS restores a NameNode with information from the Secondary NameNode, which ensures fault-tolerance of the NameNode. HDFS has a built-in balancer feature, which ensures uniform data distribution across the cluster, and re-replication of missing blocks to maintain the correct number of replications.

4.2 NoSQL Databases

Conventionally, Relational Database Management Systems (RDBMS) are used to manage large datasets and handle tons of requests securely and reliably. Built-in features, such as data integrity, security, fault-tolerance, and ACID (atomicity, consistency, isolation, and durability) have made RDBMS a go-to data management technology for organizations and enterprises. In spite of RDBMS' advantages, it is either not viable or is too expensive for applications that deal with Big data. This has made organizations to adopt a special type of database called "NoSQL" (Not an SQL), which means database systems that do not employ traditional "SQL" or adopt the constraints of the relational database model. NoSQL databases cannot provide all strong built-in features of RDBMS. Instead, they are more focused on faster read/write access to support ever-growing data.

---

[6]http://hadoop.apache.org.

According to December 2014 statistics from Facebook [16], it has 890 Million average daily active users sharing billions of messages and posts every day. In order to handle huge volumes and a variety of data, Facebook uses a Key-Value database system with memory cache technology that can handle billions of read/write requests. At any given point in time, it can efficiently store and access trillions of items. Such operations are very expensive in relational database management systems.

Scalability is another feature in NoSQL databases, attracting large number of organizations. NoSQL databases are able to distribute data among different nodes within a cluster or across different clusters. This helps to avoid capital expenditure on specialized systems, since clusters can be built with commodity computers.

Unlike relational databases, NoSQL systems have not been standardized and features vary from one system to another. Many NoSQL databases trade-off ACID properties in favor of high performance, scalability, and faster store and retrieve operations. Enumerations of such NoSQL databases tend to vary, but they are typically categorized as Key-Value databases, Document databases, Wide Column databases, and Graph databases. Figure 2 shows a hierarchical view of NoSQL types, with two examples of each type.

4.2.1 Key-Value Database

As the name suggests, Key-Value databases store data as Key-Value pairs, which makes them schema-free systems. In most of Key-Value databases, the key is functionally generated by the system, while the value can be of any data type from a character to a large binary object. Keys are typically stored in hash tables by hashing each key to a unique index.

All the keys are logically grouped, eventhough data values are not physically grouped. The logical group is referred to as a '*bucket*'. Data can only be accessed with both a bucket and a key value because the unique index is hashed using the

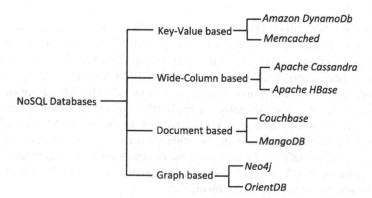

Fig. 2 Categorization of NoSQL databases: The first level in the hierarchy is the categorization of NoSQL. Second level, provides examples for each NoSQL database type

bucket and key value. The indexing mechanism increases the performance of storing, retrieving, and querying large datasets.

There are more than 40 Key-Value systems available with either commercial or open source licenses. Amazon's DynamoDB,[7] which is a commercial data storage system, and open source systems like Memcached,[8] Riak,[9] and Redis[10] are most popular examples of Key-Value database systems available.These systems differ widely in functionality and performance.

Key-Value databases are appropriate for applications that require one to store or cache unstructured data for frequent and long-term usages, such as chat applications, and social networks. Key-Value databases can also be used in applications that require real-time responses that need to store and retrieve data using primary keys, and do not need complex queries. In consumer-faced web applications with high traffic, Key-Value systems can efficiently manage sessions, configurations, and personal preferences.

4.2.2 Wide Column Database

A column-based NoSQL database management system is an advancement over a Key-Value system and is referred to as a Wide Column or column-family database. Unlike the conventional row-centric relational systems [22], Wide Column databases are column centric. In row-centric RDBMS, different rows are physically stored in different places. In contrast, column-centric NoSQL databases store all corresponding data in continuous disk blocks, which speeds up column-centric operations, such as aggregation operations. Eventhough Wide Column is an advancement over Key-Value systems, it still uses Key-Value storage in a hierarchical pattern.

In a Wide-Column NoSQL database, data are stored as name and value pairs, rather than as rows, which are known as columns. Logical grouping of columns is named as column-family. Usually the name of a column is a string, but the value can be of any data type and size (character or large binary file). Each column contains timestamp information along with a unique name and value. This timestamp is helpful to keep track of versions of that column. In a Wide-Column database, the schema can be changed at any time by simply adding new columns to column-families. All these flexibilities in the column-based NoSQL Systems are appropriate to store sparse, distributed, multidimensional, or heterogeneous data. A Wide Column database is appropriate for highly scalable applications, which require built-in versioning and high-speed read/write operations. Apache Cassandra[11] (Originated by Facebook) and Apache HBase[12] are the most widely used Wide Column databases.

---

[7]http://aws.amazon.com/dynamodb/.

[8]http://memcached.org/.

[9]http://basho.com/riak/.

[10]http://redis.io/.

[11]http://cassandra.apache.org/.

[12]http://hbase.apache.org/.

4.2.3 Document Database

A Document database works in a similar way as Wide Column databases, except that it has more complex and deeper nesting format. It also follows the Key-Value storage paradigm. However, every value is stored as a document in JSON,[13] XML[14] or other commonly used formats. Unlike Wide Column databases, the structure of each record in a Document database can vary from other records. In Document databases, a new field can be added at anytime without worrying about the schema. Because data/value is stored as a document, it is easier to distribute and maintain data locality. One of the disadvantages of a Document database is that it needs to load a lot of data, even to update a single value in a record. Document databases have built-in approach of updating a document, while retaining all old versions of the document. Most Document database systems use secondary indexing [26] to index values and documents in order to obtain faster data access and to support query mechanisms. Some of the database systems offer full-text search libraries and services for real-time responses.

One of the major functional advantages of document databases is the way it interfaces with applications. Most of the document database systems use JavaScript (JS) as a native scripting language because it stores data in JS friendly JSON format. Features such as JS support, ability to access documents by unique URLs, and ability to organize and store unstructured data efficiently, make Document databases popular in web-based applications. Documents databases serve a wide range of web applications, including blog engines, mobile web applications, chat applications, and social media clients.

Couchbase[15] and MongoDB[16] are among popular document-style databases. There are over 30 document databases. Most of these systems differ in the way data are distributed (both partition and replications), and in the way a client accesses the system. Some systems can even support transactions [23].

4.2.4 Graph Databases

All NoSQL databases partition or distribute data in such a way that all the data are available in one place for any given operation. However, they fail to consider the relationship between different items of information. Additionally, most of these systems are capable of performing only one-dimensional aggregation at a time.

---

[13]http://json.org.
[14]http://www.w3.org/TR/2006/REC-xml11-20060816/.
[15]http://couchbase.com/.
[16]http://mangodb.org/.

A Graph database is a special type of database that is ideal for storing and handling relationship between data. As the name implies Graph databases use a graph data model. The vertices of a graph represent entities in the data and the edges represent relationships between entities. Graph data model, perfectly fits for scaling out and distributing across different nodes. Common analytical queries in Graph databases include finding the shortest path between two vertices, identifying clusters, and community detection.

Social graphs, World Wide Web, and the Semantic Web are few well-known use cases for graph data models and Graph databases. In a social graph, entities like friends, followers, endorsements, messages, and responses are accommodated in a graph database, along with relationships between them. In addition to maintaining relationships, Graph databases make it easy to add new edges or remove existing edges. Graph databases also support the exploration of time-evolving graphs by keeping track of changes in properties of edges and vertices using time stamping.

There are over 30 graph database systems. Neo4j[17] and Orient DB[18] are popular examples of graph-based systems. Graph databases found their way into different domains, such as social media analysis (e.g., finding most influential people), e-commerce (e.g., developing recommendations system), and biomedicine (e.g., to analyze and predict interactions between proteins). Graph databases also serve in several industries, including airlines, freight companies, healthcare, retail, gaming, and oil and gas exploration.

4.2.5 Cloud-Based NoSQL Database Services

Amazon DynamoDB: DynamoDB[19] is a reliable and fully managed NoSQL data service, which is a part of Amazon Web Services (AWS). It is a Key-Value database that provides a schema-free architecture to support ever-growing Big data in organizations and real-time web applications. DynamoDB is well optimized to handle huge volume of data with high efficiency and throughput. This system can scale and distribute data, virtually, without any limit. DynamoDB partitions data using a hashing method and replicates data three times and distributes them among data centers in different regions in order to enable high availability and fault tolerance. DynamoDB automatically partitions and re-partitions data depending on data throughput and volume demands. DynamoDB is able to handle unpredictable workloads and high volume demands efficiently and automatically.

DynamoDB offers eventual and strong consistency for read operations. Eventual consistency does not always guarantee that a data read is the latest written version of the data, but significantly increases the read throughput. Strong consistency guarantees that values read are the latest values after all write operations. DynamoDB allows the user to specify a consistency level for every read operation. DynamoDB

---

[17]http://neo4j.org/.

[18]http://www.orientechnologies.com/orientdb/.

[19]http://aws.amazon.com/dynamodb/.

also offers secondary indexing (i.e., local secondary and global secondary), along with the indexing of the primary key for faster retrieval.

DynamoDB is a cost efficient and highly scalable NoSQL database service from Amazon. It offers benefits such as reduced administrative supervision, virtually unlimited data throughput, and the handling of all the workloads seamlessly.

Google BigQuery: Google uses massively parallel query system called as 'Dremel' to query very large datasets in seconds. According to [50], Dremel can scan 35 billion rows in ten seconds even without indexing. This is significantly more efficient than querying a Relational DBMS. For example, on Wikipedia dataset with 314 million rows, Dremel took 10 seconds to execute regular expression query to find the number of articles in Wikipedia that include a numerical character in the title [50]. Google is using Dremel in web crawling, Android Market, Maps, and Books services.

Google brought core features of this massive querying system to consumers as a cloud-based service called 'BigQuery'.[20] Third party consumers can access BigQuery through either a web-based user interface, command-line or through their own applications using the REST API. In order to use BigQuery features, data has to be transferred into the Google Cloud storage in JSON encoding. The BigQuery also returns results in JSON format.

Along with an interactive and fast query system, Google cloud platform also provides automatic data replication, on-demand scalability, and handles software and hardware failure without administrative burdens. In 2014, using BigQuery, scanning one terabyte of data only cost $5, with additional cost for storage.[21]

Windows Azure Tables: Windows Azure Tables[22] is a NoSQL database technology with a Key-Value store on the Windows Azure platform. Azure Tables also provides, virtually, unlimited storage of data. Azure Tables is highly scalable and supports automatic partitioning. This database system distributes data across multiple machines efficiently to provide high data throughput and to support higher workloads. Azure Tables storage provides the user with options to select a Partition-Key and a Row-Key upfront, which may later be used for automatic data partitioning. Azure Tables follows only the strong consistency data model for reading data. Azure Tables replicates data three times among data centers in the same region and additional three times in other regions to provide a high degree of fault-tolerance.

Azure Tables is a storage service for applications with huge volume of data, and needs schema-free NoSQL databases. Azure Tables uses primary key alone and it does not support secondary indexes. Azure Tables provides the REST-based API to interact with its services.

---

[20]http://cloud.google.com/bigquery/.
[21]https://cloud.google.com/bigquery/pricing.
[22]http://azure.microsoft.com/.

5 Examples of Massive Data Applications

In this section, a detailed discussion of solutions proposed by our research team for two real-world Big data problems are presented.

5.1 Recommendations in e-Commerce

Recommender systems are gaining wide popularity in e-commerce, as they are becoming major drivers of incremental business value and user satisfaction [29, 31]. In this section, we will describe the architecture behind a recommendation engine for eBay, a large open marketplace [28]. In an e-commerce system, there are two major kinds of recommendation scenarios: pre-purchase and post-purchase.

In the pre-purchase scenario, the system recommends items that are good alternatives for the item the user is viewing. In the post-purchase scenario, the recommendation system recommends items complementary or related to an item, which the user has bought recently.

5.1.1 Architecture

The architecture of the recommendation system, as illustrated in Fig. 3, consists of the Data Store, the Real-time Performance System, and the Offline Model Generation System. The Data Store holds the changes to website data as well as models learned.

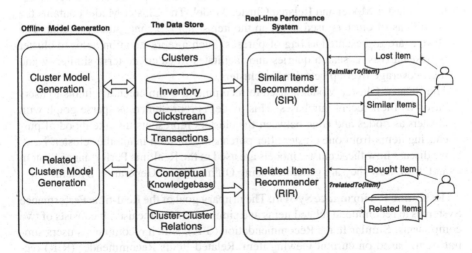

Fig. 3 The recommendation system architecture with three major groups: The Offline Modeling System; The Data Store; The Real-time Performance System

The Real-time Performance System is responsible for recommending items using a session state of the user and contents from Data Store. The Offline Model Generation System is responsible for building models using computationally intensive offline analyze. Next, we present a detailed discussion about these components.

Data Store: The Data Store provides data services to both the Offline Model Generation and the Real-time Performance components. It provides customized versions of similar services to each of these components. For example, we consider a service that provides access to item inventory data. The Offline Modeling component has access to longitudinal information of items in the inventory, but not an efficient way of keyword search. On the other hand, the Real-time Performance System does not have access to longitudinal information, but it can efficiently search for item properties in the current inventory. Two types of data sources are used by our system: Input Information sources and Output Cluster models.

- *Input Information Sources*:
  The Data Store is designed to handle continuous data sources such as users' actions and corresponding state changes of a website. At the same time, it also stores models, which are generated by the Offline Model Generation System. The data in the Data Store can be broadly categorized into inventory data, clickstream data, transaction data, and conceptual knowledge base. The inventory data contains the items and their properties. Clickstream data includes the raw data about the users' actions with dynamic state of the website. Even though the purchasing history can be recreated from clickstream data, it is stored separately as transaction data for efficient access. Conceptual knowledge base includes ontology-based hierarchical organization of items, referred to as the category tree, lexical knowledge source, and term dictionary of category-wise important terms/phrases.
- *Output Cluster Model*: The Data Store contains two types of knowledge structures: Cluster Model and Related Cluster Model. The Cluster Model contains the definitions of clusters used to group the items that are conceptually similar. The clusters are represented as bag-of-phrases. Such a representation helps to cluster representatives as search queries and facilitates to calculate term similarity and item-coverage overlap between the clusters.
  The Related Cluster Model is used to recommend complementary items to users based on their recent purchases. This model is represented as sparse graph with clusters as nodes and edge between the clusters represents the likelihood of purchasing items from one cluster after purchasing an item in another cluster. Next, we discuss how these cluster models are used in the Realtime Performance System and, then, how they are generated using Offline Model Generation System.

Real-time Performance System: The primary goal of the Real-time Performance System is to recommend related items and similar items to the user. It consists of two components, Similar Items Recommendation (SIR), which recommends users similar items based on current viewing item. Related Items Recommender (RIR) recommends users the related items based on their recent purchases. Real-time Performance System is essential to generate the recommendations in real-time to honor the dynamic user actions. To achieve this performance, any computationally intensive decision process is compiled to offline model. It is required to indexed data source such that it can be queried efficiently and to limit the computation after retrieving.

The cluster assignment service generates normalized versions of a cluster as a Lucene[23] index. This service performs similar normalization on clusters and input item's title and its static properties, to generate the best matching clusters. The SIR and RIR systems use the matching clusters differently. SIR selects the few best items from the matching clusters as its recommendations. However, RIR picks one item per query it has constructed to ensure the returned recommendations relates to the seeded item in a different way.

Offline Model Generation:

- *Clusters Generation*: The inventory size of an online marketplace ranges in the hundreds of millions of items and these items are transient, i.e., covering a broad spectrum of categories. In order to cluster such a large scale and diverse inventory, the system uses distributed clustering approach on a Hadoop Map-Reduce cluster, instead of a global clustering approach.
- *Cluster-Cluster Relations Generation*: An item-to-item co-purchase matrix is generated using the purchase history of users from the transactional data set. Hadoop Map-Reduce clusters are employed to compute Cluster-related cluster pairs from the item-to-item co-purchase matrix.

5.1.2 Experimental Results

We conducted A/B tests to compare the performance of our Similar and Related Items Recommender systems described in this section over the legacy recommendation system developed by Chen & Canny [11]. The legacy system clusters the items using generative clustering and later it uses a probabilistic model to learn relationship patterns from the transaction data. One of the main differences is the way these two recommendation system generate the clusters. The legacy system uses item data (auction title, description, price), whereas our system uses user queries to generate clusters.

A test was conducted on Closed View Item Page (CVIP) in eBay to compare our Similar Items Recommender algorithm with the legacy algorithm. CVIP is a page that is used to engage a user by recommending similar items after an unsuccessful bid. We also conducted a test to compare our Related Items Recommender with legacy algorithm [11]. Both the test results show significant improvement in user engagement and site-wide business metrics with 90 % confidence. As we are not permitted to publish actual figures representing system performances, we are reporting relative statistics. Relative improvements in user engagement (Click Through Rate) with our SIR and RIR, over legacy algorithms, are 38.18 % and 10.5 %, respectively.

---

[23]http://lucene.apache.org/.

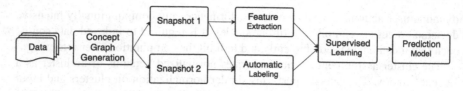

Fig. 4 Implementation flow diagram for supervised link prediction in biomedical literature

5.2 Link Prediction in Biomedical Literature

Predicting the likelihood of two nodes associating in the future, which do not have direct association between them in the current timestep, is known as the link prediction problem. Link prediction is widely used in social network analysis. Link prediction has wide range of applications such as identifying missing information, identifying spurious interactions, and studying the evolution of the network. In e-commerce, link prediction is used for building recommendation systems, and in bioinformatics, it is used to predict protein–protein interactions.

Katukuri et al. [30], proposed a supervised link prediction method, to predict unknown association of medical concepts using biomedical publication information from Medline.[24] Medline is a National Institute of Health (NIH)'s citation database with more than 21 million publication citations. Figure 4 illustrates different stages in the proposed supervised link prediction approach. A temporal concept network is generated using relevant medical concepts extracted from publications. In the concept network, each node represents a medical concept and an edge between two nodes represents relationship that two medical concepts co-occurred at least in one publication. Document frequency of a given concept is a weight of node and co-occurrence frequency of two concepts is edge weight. Now, link prediction problem is formulated as a process of identifying whether a pair of concepts, which are not directly connected in the current duration concept network, will be connected directly in the future.

This link prediction problem is formulated as a supervised classification task. Training data is automatically labeled by comparing concept network snapshots of two consecutive time periods. This automatic labeling approach helps to avoid need for domain experts.

In automatic labeling method, concept pairs, which are not directly connected in the first snapshot, are labeled based on its possible connection strength in the second snapshot. Connections strength is categorized as follows (S is edge weight in second snapshot, and minimum_support and margin (ranges between 0 to 1) are user-defined values):

- Connection as strong: $S \geq$ minimum_support
- Connection as emerging: *margin* $\times$ *minimum_support* $\leq S <$ minimum_support.

---

[24]http://www.ncbi.nlm.nih.gov/pubmed/.

- Connection as weak: S < *margin* × *minimum_support*.
- No connection: S=0.

Given a pair of nodes that has no direct connection in first snapshot is assigned with positive class label if this pair is strongly connected in the second snapshot and is assigned negative class label if it has weak connection or no connection the second snapshot, and the pairs with intermediate values of strength are labeled as emerging.

For each of labeled concept pairs, a set of topological features (random walk based and neighborhood-based) is extracted from the first snapshot of the concept network. The topological feature set also includes Common neighbors, Adamic/Adar, Jaccard Co-efficient, and Preferential Attachment [4]. Along with topological features, semantically enriched features, like Semantic CFEC [30] are extracted. Combining labeled concept pairs with the corresponding feature set generates the training instances. Supervised classification algorithms, such as SVM, and C4.5 decision tree are used to generate prediction models. The classification accuracy of prediction models is calculated using cross-validation, which is on average 72 % [30].

Implementing such a computationally intensive phase to extract features needed for generating the predictive model on massive data needs large computational resources. For example, a snapshot of the concept network for years 1991–2010 has nearly 0.2 million nodes and nearly 44 million edges. Processing millions of publications to extract medical concepts, generating class labels, and extracting features from large well-connected network is computationally heavy. To handle such computations on large graphs, the prediction system is developed by using the MapReduce framework and a Hadoop cluster environment. We implemented the MapReduce functions to extract the medical concepts from millions of publications in Medline dataset, to generate the labeled data, and to extract structural features from the large concept graph.

Implementing such a graph computation method on MapReduce framework has its own limitations. One of the drawbacks of MapReduce framework is its inability to retain the state of a graph across multiple iterations [44]. One approach to retain the state in a MapReduce pipeline is by explicitly writing the graph to disk after one iteration and reading it back from disk in the next iteration. This approach proves to be inefficient due to the huge increase in I/O operation and bandwidth [44].

Google proposed a Bulk Synchronous Parallel processing model called Pregel [37], which is a message passing model. Unlike MapReduce framework, this model helps by retaining the graph state across the iterations. Apache Giraph is open source alternative to Pregel, which is built on top of MapReduce framework. In such distributed graph systems, a graph is partitioned and distributed among different cluster nodes. Each vertex has information about itself along with its neighbors. In our link prediction, features like the Jaccard Coefficient, can be extracted in parallel since such calculations depend only on information local to each vertex. However, other features, like the Semantic CFEC, need to be calculated by exploring all the paths between given pair of nodes, which can be formulated as an all-pairs-path problem. There are several frameworks [36, 37] that can calculate the all-pairs-path problem by passing information through edges, but between just a pair of nodes at a time.

However, these frameworks cannot support the operation of finding all paths between all pairs in parallel. In our case, there is a need to extract such features for millions of concept pairs. To the best of our knowledge, there is no algorithm or framework that can support such a problem to run in a distributed environment. This is one of the open computational challenges in graph analytics that needs to be investigated by the research community.

6 Current Issues of Big Data Analytics

In this section, we discuss several open challenges relating to computation, storage, and security in Big data analytics.

6.1 Data Locality

One prominent aspect in efficient Big data processing is the ability to access data without a significant latency. Given the transfer-prohibitive volume of Big data, accessing data with low latency can be accomplished only through data locality. In fact, data movement is possible in computations with moderate to medium volume of data where the data transfer to processing time ratio is low. However, this is not the case for Big data analytics applications. The alternative approach to alleviate the data transfer problem is moving the computation to where the data resides. Thus, efficient data management policies are required in the Big data analytics platforms to consider issues such as maximizing data locality and minimizing data migration (i.e., data transfer) between cloud data centers [63].

One of the key features of the Hadoop framework is its ability to take the effects of data locality into account. In Hadoop, the JobTracker component tries to allocate jobs to nodes where the data exists. Nonetheless, there are cases in which all the nodes that host a particular data node are overloaded. In this situation, JobTracker has to schedule the jobs on machines that do not have the data.

To expedite data processing, Spark keeps the data in main memory, instead of on disk. Spark's data locality policy is similar to Hadoop. However, in Spark, the Reduce tasks are allocated to machines where the largest Map outputs are generated. This reduces data movement across the cluster and improves the data locality further.

6.2 Fault-Tolerance of Big Data Applications

In long-running, Big data analytics applications, machine failure is inevitable. Both transient (i.e., fail-recovery) and permanent (i.e., fail-stop) failures can occur during the execution of such applications [32]. Google reports experiencing on average, five machine crashes during a MapReduce job in March 2006 [12] and at a minimum one disk failure in each execution of a MapReduce job with 4000 tasks. Because of the criticality of failure, any resource allocation method for Big data jobs should be fault-tolerant.

MapReduce was originally designed to be robust against faults that commonly happen at large-scale resource providers with many computers and devices such as network switches and routers. For instance, reports show that during the first year of a cluster operation at Google there were 1000 individual machine failures and thousands of hard-drive failures.

MapReduce uses logs to tolerate faults. For this purpose, the output of Map and Reduce phases create logs on the disk [39]. In the event that a Map task fails, it is re-executed with the same partition of data. In case of failure in Reducer, the key/value pairs for that failed Reducer are regenerated.

6.3   Replication in Big Data

Big data applications either do not replicate the data or do it automatically through a distributed file system (DFS). Without replication, the failure of a server storing the data causes the re-execution of the affected tasks. Although the replication approach provides more fault-tolerance, it is not efficient due to network overhead and increasing the execution time of the job.

Hadoop platform provides user a static replication option to determine the number of times a data block should replicate within the cluster. Such a static replication approach adds significant storage overhead and slows down the job execution. A solution to handle this problem is the dynamic replication that regulates the replication rate based on the usage rate of the data. Dynamic replication approaches help to utilize the storage and processing resources efficiently [64].

Cost-effective incremental replication [34] is a method, for cloud-based jobs, that is capable of predicting when a job needs replication. There are several other data replication schemes for Big data applications on clouds. In this section, we discuss four major replication schemes namely, Synchronous and Asynchronous replication, Rack-level replication, and Selective replication. These replication schemes can be applied at different stages of the data cycle.

Synchronous data replication scheme (e.g., HDFS) ensures data consistency through blocking producer tasks in a job until replication finishes. Even though Synchronous data replication yields high consistency, it introduces latency, which affects the performance of producer tasks. In Asynchronous data replication scheme [32], a producer proceeds regardless of the completion of a producer task on a replica of the same block. Such a nonblocking nature of this scheme improves the performance of the producer. But consistency of Asynchronous replication is not as precise as the Synchronous replication. When Asynchronous replication is used in the Hadoop framework, Map and Reduce tasks can continue concurrently.

Rack-level data replication scheme ensures that all the data replicas occur on the same rack in a data center. In fact, in data centers, servers are structured in racks with a hierarchical topology. In a two-level architecture, the central switch can become the bottleneck as many rack switches share it. One instance of bandwidth bottleneck is in the Shuffling step of MapReduce. In this case, the central switch becomes over utilized, whereas rack-level switches are under-utilized. Using the Rack-level replication helps reduce the traffic that goes through the central switch. However, this schema cannot tolerate rack-level failures. But recent studies suggest, the rack-level failures are uncommon, which justifies the adoption of Rack-level replication.

In *Selective data replication*, intermediate data generated by the Big data application are replicated on the same server where they were generated. For example, in the case of Map phase failures in a chained MapReduce job, the affected Map task can be restarted directly, if the intermediate data from previous Reduce tasks were available on the same machine. The Selective replication scheme reduces the need for replication in the Map phase. However, it is not effective in Reduce phase, since the Reduce data are mostly consumed locally.

Data replication on distributed file systems is costly due to disk I/O operations, network bandwidth, and serialization overhead. These overheads can potentially dominate the job execution time [65]. Pregel [37], a framework for iterative graph computation, stores the intermediate data in memory to reduce these overheads. Spark [66] framework uses a parallel data structures known as Resilient Distributed Datasets (RDDs) [65] to store intermediate data in memory and manipulate them using various operators. They also control the partitioning of the data to optimize the data placement.

6.4 Big Data Security

In spite of the advantages offered by Big data analytics on clouds and the idea of Analytics as a Service, there is an increasing concern over the confidentiality of the Big data in these environments [6]. This concern is more serious as increasing amount of confidential user data are migrated to the cloud for processing. Genome sequences, health information, and feeds from social networks are few instances of such data.

A proven solution to the confidentiality concerns of sensitive data on cloud is to employ user-side cryptographic techniques for securing the data [6]. However, such techniques limit the cloud-based Big data analytics in several aspects. One limitation is that the cryptographic techniques usually are not transparent to end users. More importantly, these techniques restrict functionalities, such as searching and processing, that can be performed on the users' data. Numerous research works are being undertaken to address these limitations and enable seamless Big encrypted data analytics on the cloud. However, all of these efforts are still in their infancy and not applicable to Big data scale processing.

Another approach to increase data confidentiality is to utilize multiple cloud storage units simultaneously [60]. In this approach, user data are sharded based on a user-side hashing function, and then the data for each cloud is encrypted and uploaded across multiple clouds. It is noteworthy that sharding and distributing of the data are achieved based on some lightweight user-side processing. Therefore, the control and distribution of the data is determined merely at the user-side. Although such sharding approach seems interesting for Big data analytics, challenges, such as processing sharded data across multiple clouds, still remains unsolved.

Hybrid clouds have proven to be helpful in increasing the security of Big data analytics. In particular, they can be useful for cloud-based Big data analytics where a portion of the data is sensitive and needs specific trusted resources for execution. One approach is to label the data and treat them differently based on their labels [67]. As such, nonsensitive data are pushed to a public cloud for processing and sensitive data are processed in a private cloud. The coordination is accomplished through a scheduler placed within local resources that determines where a data should be processed depending on its label.

6.5 Data Heterogeneity

One of the major challenges researchers are facing is "How to integrate all the data from different sources to maximize the value of data." In the World Wide Web (WWW), there is a huge amount of data created by social network sites, blogs, and websites. However, every source is different in data structure, semantics, and format. Structure of data from these sources varies from well-structured data (e.g., databases) to unstructured data (e.g., heterogeneous documents).

Singh et al. [53], developed a framework to detect situations (such as epidemics, traffic jams) by combining information from different streams like Twitter, Google Insights, and satellite imagery. In this framework, heterogeneous real-time data streams are combined by converting selected attributes and unified across all streams. There are several other proposed frameworks that can combine different data sources for various chosen domain-specific applications. Most of these solutions use a semantics-based approach. Ontology matching is a popular semantics-based method, which finds the similarity between the ontologies of different sources.

Ontology is a vocabulary and description of a concept and its relationship with others in the respective domain. In the Web example, ontology is used to transform unstructured or partially structured data from different sources. Most of them are human readable format (e.g., HTML) and are hard for the machine to understand. One of the most successful ontology integration projects is Wikipedia, which is essentially integrated with human intervention. Semantic Web tools are used to convert the unstructured Web to a machine understandable structure. Semantic Web adds ontology constructs into web pages and enables machines to understand the contents of a webpage. It helps to automate integration of heterogeneous data from different sources on the Web using ontology matching.

7   Summary and Discussion

Size of data in the digital universe is almost doubling every year and has reached to a stage that they cannot be processed by the conventional programming, computing, storage, visualization, and analytical tools. In this study, we reviewed different types of analytic needs arising in research and industry. We broadly categorized the current analytical applications as descriptive, predictive, and prescriptive and identified several real-world applications of each type.Then, we provided an overview of the state-of-the-art on platforms, tools, and use cases for massive data analytics.

In particular, we discussed that cloud services are helpful platforms in alleviating many of the massive data processing challenges. MapReduce compute resources, NoSQL databases, virtually unlimited storages, and customized filesystems, amongst many others, are useful cloud services for massive data analytics.

We provided two use cases that were investigated within our research team. The first one, recommendation in e-commerce applications, consists of a number of components that can be partitioned into three major groups: *The Data Store* that contains data about the active and temporarily changing state of an e-commerce web site; *The Real-time Performance System* that generates recommendations in real-time based on the information in the Data Store; and the *offline model generation* that conducts computationally intensive offline analyses. The Real-time Performance System consists of two components, similar items recommender (SIR) and related items recommender (RIR). Both of these components take a seed item as input, and return a set of items that are similar or related to that seed item.

The second use case addresses the problem of link prediction that proposes associations between medical concepts that did not exist in earlier published works. In this project, we model biomedical literature as a concept network, where each node represents a biomedical concept that belongs to a certain semantic type, and each edge represents a relationship between two concepts. Each edge is attached with a weight that reflects the significance of the edge. Based on the constructed massive graph, a machine-learning engine is deployed to predict the possible connection between two indirectly connected concepts.

In the course of our research on massive data analytics tools and projects, we have learnt key lessons and identified open challenges that have to be addressed by researchers to further advance efficient massive data analytics. Below, we highlight some of the lessons and challenges:

- In many analytical applications (e.g., recommendation system in e-commerce), even with availability of state-of-the-art Big data technologies, treating customer data as a data stream is not yet viable. Therefore, some steps (e.g., model building in recommendation systems), have to be performed offline.
- It is difficult, if not impossible, to come up with a generic framework for various types of analytics. For instance, in the recommendation system, which is an example of predictive data analytics in e-commerce, there are many subtle nuances. Thus, a specific architecture is required based on the merits of each application. Accordingly, in the eBay application, we noticed that Related Items Recommen-

dation (RIR) needs a different architecture compared to Similar Items Recommendation (SIR).

- Many Big data analytics (e.g., biomedical link prediction) process massive graphs as their underlying structure. Distributed graph techniques need to be in place for efficient and timely processing of such structures. However, to the best our knowledge, there is not yet a comprehensive distributed graph analytic framework that can support all conventional graph operations (e.g., path-based processing in distributed graphs).
- Data locality and replication management policies ought to be cleverly integrated to provide robust and fault-tolerant massive data analytics.
- As massive data are generally produced from a great variety of sources, novel, semantics-based solutions should be developed to efficiently support data heterogeneity.

References

1. 5 must-have lessons from the 2014 holiday season. http://www.experian.com/blogs/marketing-forward/2015/01/14/five-lessons-from-the-2014-holiday-season/. Accessed 6 March 2015
2. 6 uses of big data for online retailers. http://www.practicalecommerce.com/articles/3960-6-Uses-of-Big-Data-for-Online-Retailers. Accessed 28 Feb 2015
3. Abbasi A, Albrecht C, Vance A, Hansen J (2012) Metafraud: a meta-learning framework for detecting financial fraud. MIS Q 36(4):1293–1327
4. Al Hasan M, Zaki MJ (2011) A survey of link prediction in social networks. In: Aggarwal CC (ed) Social network data analytics, pp 243–275. Springer US
5. Alaçam O, Dalcı D (2009) A usability study of webmaps with eye tracking tool: The effects of iconic representation of information. In: Proceedings of the 13th international conference on human-computer interaction. Part I: new trends, pp 12–21. Springer
6. Amini Salehi M, Caldwell T, Fernandez A, Mickiewicz E, Redberg D, Rozier EWD, Zonouz S (2014) RESeED: regular expression search over encrypted data in the cloud. In: Proceedings of the 7th IEEE Cloud conference, Cloud '14, pp 673–680
7. Assunção MD, Calheiros RN, Bianchi S, Netto MAS, Buyya R (2014) Big data computing and clouds: Trends and future directions. J Parallel Distrib Comput
8. Australian square kilometer array pathfinder radio telescope. http://www.atnf.csiro.au/projects/askap/index.html. Accessed 28 Feb 2015
9. Big data and content analytics: measuring the ROI. http://www.aiim.org/Research-and-Publications/Research/Industry-Watch/Big-Data-2013. Accessed 28 Feb 2015
10. Buckinx W, Verstraeten G, Van den Poel D (2007) Predicting customer loyalty using the internal transactional database. Expert Syst Appl 32(1):125–134
11. Chen Y, Canny JF (2011) Recommending ephemeral items at web scale. In: Proceedings of the 34th international ACM SIGIR conference on research and development in information retrieval, pp 1013–1022. ACM
12. Dean J (2006) Experiences with MapReduce, an abstraction for large-scale computation. In: Proceedings of the 15th international conference on parallel architectures and compilation techniques, PACT '06
13. Dean J, Ghemawat S (2008) MapReduce: Simplified data processing on large clusters. Commun ACM 51(1):107–113
14. Descriptive, predictive, prescriptive: transforming asset and facilities management with analytics (2013)

15. Enhancing exploration and production with big data in oil & gas. http://www-01.ibm.com/software/data/bigdata/industry-oil.html. Accessed 28 Feb 2015
16. Facebook. http://newsroom.fb.com/company-info/. Accessed 14 March 2015
17. Farris A (2012) How big data is changing the oil & gas industry. Analyt Mag
18. Gartner survey reveals that 73 percent of organizations have invested or plan to invest in big data in the next two years. http://www.gartner.com/newsroom/id/2848718. Accessed 28 Feb 2015
19. Gartner taps predictive analytics as next big business intelligence trend. http://www.enterpriseappstoday.com/business-intelligence/gartner-taps-predictive-analytics-as-next-big-business-intelligence-trend.html. Accessed 28 Feb 2015
20. Ghemawat S, Gobioff H, Leung ST (2003) The google file system. In: Proceedings of the 19th ACM symposium on operating systems principles, SOSP '03, pp 29–43
21. Gudivada VN, Baeza-Yates R, Raghavan VV (2015) Big data: promises and problems. Computer 3:20–23
22. Gudivada VN, Rao D, Raghavan VV (2014) NoSQL systems for big data management. In: IEEE World congress on Services (SERVICES), 2014, pp 190–197. IEEE
23. Gudivada VN, Rao D, Raghavan VV (2014) Renaissance in data management systems: SQL, NoSQL, and NewSQL. IEEE Computer (in Press)
24. HPCC vs Hadoop. http://hpccsystems.com/Why-HPCC/HPCC-vs-Hadoop. Accessed 14 March 2015
25. IBM netfinity predictive failure analysis. http://ps-2.kev009.com/pccbbs/pc_servers/pfaf.pdf. Accessed 14 March 2015
26. Indrawan-Santiago M (2012) Database research: are we at a crossroad? reflection on NoSQL. In: 2012 15th International conference on network-based information systems (NBiS), pp 45–51
27. Instagram. https://instagram.com/press/. Accessed 28 Feb 2015
28. Jayasimha K, Rajyashree M, Tolga K (2013) Large-scale recommendations in a dynamic marketplace. In Workshop on large scale recommendation systems at RecSys 13:
29. Jayasimha K, Rajyashree M, Tolga K (2015) Subjective similarity: personalizing alternative item recommendations. In: WWW workshop: Ad targeting at scale
30. Katukuri JR, Xie Y, Raghavan VV, Gupta A (2012) Hypotheses generation as supervised link discovery with automated class labeling on large-scale biomedical concept networks. BMC Genom 13(Suppl 3):S5
31. Katukuri J, Konik ,T Mukherjee R, Kolay S (2014) Recommending similar items in large-scale online marketplaces. In: 2014 IEEE International conference on Big Data (Big Data), pp 868–876. IEEE
32. Ko SY, Hoque I, Cho B, Gupta I (2010) Making cloud intermediate data fault-tolerant. In: Proceedings of the 1st ACM symposium on cloud computing, SoCC '10, pp 181–192
33. Lam C (2010) Hadoop in action, 1st edn. Manning Publications Co., Greenwich, CT, USA
34. Li W, Yang Y, Yuan D (2011) A novel cost-effective dynamic data replication strategy for reliability in cloud data centres. In: Proceedings of the Ninth IEEE international conference on dependable, autonomic and secure computing, DASC '11, pp 496–502
35. Lohr S (2012) The age of big data. New York Times 11
36. Low Y, Gonzalez J, Kyrola A, Bickson D, Guestrin C, Hellerstein JM (2010) Graphlab: a new framework for parallel machine learning. arxiv preprint. arXiv:1006.4990
37. Malewicz G, Austern MH, Bik AJC, Dehnert JC, Horn I, Leiser N, Czajkowski G (2010) Pregel: a system for large-scale graph processing. In: Proceedings of the ACM SIGMOD international conference on management of data, SIGMOD '10, pp 135–146
38. Manyika J, Michael C, Brad B, Jacques B, Richard D, Charles R (2011) Angela Hung Byers, and McKinsey Global Institute. The next frontier for innovation, competition, and productivity, Big data
39. Martin A, Knauth T, Creutz S, Becker D, Weigert S, Fetzer C, Brito A (2011) Low-overhead fault tolerance for high-throughput data processing systems. In: Proceedings of the 31st International conference on distributed computing systems, ICDCS '11, pp 689–699

40. Middleton AM, Bayliss DA, Halliday G (2011) ECL/HPCC: A unified approach to big data. In: Furht B, Escalante A (eds) Handbook of data intensive computing, pp 59–107. Springer, New York
41. Middleton AM (2011) Lexisnexis, and risk solutions. White Paper HPCC systems: data intensive supercomputing solutions. Solutions
42. NASA applies text analytics to airline safety. http://data-informed.com/nasa-applies-text-analytics-to-airline-safety/. Accessed 28 Feb 2015
43. New IDC worldwide big data technology and services forecast shows market expected to grow to $32.4 billion in 2017. http://www.idc.com/getdoc.jsp?containerId=prUS24542113. Accessed 28 Feb 2015
44. Processing large-scale graph data: A guide to current technology. http://www.ibm.com/developerworks/library/os-giraph/. Accessed 08 Sept 2015
45. Purdue university achieves remarkable results with big data. https://datafloq.com/read/purdue-university-achieves-remarkable-results-with/489. Accessed 28 Feb 2015
46. Reinsel R, Minton S, Turner V, Gantz JF (2014) The digital universe of opportunities: rich data and increasing value of the internet of things
47. Resources:HPCC systems. http://hpccsystems.com/resources. Accessed 14 March 2015
48. Russom P et al (2011) Big data analytics. TDWI Best Practices Report, Fourth Quarter
49. Salehi M, Buyya R (2010) Adapting market-oriented scheduling policies for cloud computing. In: Algorithms and architectures for parallel processing, vol 6081 of ICA3PP' 10. Springer, Berlin, pp 351–362
50. Sato K (2012) An inside look at google bigquery. White paper. https://cloud.google.com/files/BigQueryTechnicalWP.pdf
51. Shinnar A, Cunningham D, Saraswat V, Herta B (2012) M3r: Increased performance for in-memory hadoop jobs. Proc VLDB Endown 5(12):1736–1747
52. Shvachko K, Kuang H, Radia S, Chansler R (2010) The hadoop distributed file system. In: Proceedings of the 26th IEEE symposium on mass storage systems and technologies, MSST '10, pp 1–10
53. Singh VK, Gao M, Jain R (2012) Situation recognition: an evolving problem for heterogeneous dynamic big multimedia data. In: Proceedings of the 20th ACM international conference on multimedia, MM '12, pp 1209–1218, New York, NY, USA, 2012. ACM
54. The future of big data? three use cases of prescriptive analytics. https://datafloq.com/read/future-big-data-use-cases-prescriptive-analytics/668. Accessed 02 March 2015
55. The large Hadron collider. http://home.web.cern.ch/topics/large-hadron-collider. Accessed 28 Feb 2015
56. The oil & gas industry looks to prescriptive analytics to improve exploration and production. https://www.exelisvis.com/Home/NewsUpdates/TabId/170/ArtMID/735/ArticleID/14254/The-Oil--Gas-Industry-Looks-to-Prescriptive-Analytics-To-Improve-Exploration-and-Production.aspx. Accessed 28 Feb 2015
57. Troester M (2012) Big data meets big data analytics, p 13
58. Van den Poel D, Buckinx W (2005) Predicting online-purchasing behaviour. Eur J Oper Res 166(2):557–575
59. VC funding trends in big data (IDC report). http://www.experfy.com/blog/vc-funding-trends-big-data-idc-report/. Accessed 28 Feb 2015
60. Wang J, Gong W, Varman P, Xie C (2012) Reducing storage overhead with small write bottleneck avoiding in cloud raid system. In: Proceedings of the 2012 ACM/IEEE 13th international conference on grid computing, GRID '12, pp 174–183, Washington, DC, USA, 2012. IEEE Computer Society
61. Wolpin S (2006) An exploratory study of an intranet dashboard in a multi-state healthcare system
62. Youtube statistics. http://www.youtube.com/yt/press/statistics.html. Accessed 28 Feb 2015
63. Yuan D, Yang Y, Liu X, Chen J (2010) A data placement strategy in scientific cloud workflows. Fut Gen Comput Syst 26(8):1200–1214

64. Yuan D, Cui L, Liu X (2014) Cloud data management for scientific workflows: research issues, methodologies, and state-of-the-art. In: 10th International conference on semantics, knowledge and grids (SKG), pp 21–28

65. Zaharia M, Chowdhury M, Das T, Dave A, Ma J, McCauley M, Franklin MJ, Shenker S, Stoica I (2012) Resilient distributed datasets: a fault-tolerant abstraction for in-memory cluster computing. In: Proceedings of the 9th USENIX conference on networked systems design and implementation, NSDI'12, pp 2–12. USENIX Association

66. Zaharia M, Chowdhury M, Franklin MJ, Shenker S, Stoica I (2010) Spark: Cluster computing with working sets. In: Proceedings of the 2nd USENIX conference on hot topics in cloud computing, HotCloud'10, pp 10–15

67. Zhang C, Chang EC, Yap RHC (2014) Tagged-MapReduce: a general framework for secure computing with mixed-sensitivity data on hybrid clouds. In: Proceedings of 14th IEEE/ACM international symposium on cluster, cloud and grid computing, pp 31–40

Statistical Challenges with Big Data in Management Science

Arnab Laha

Abstract In the past few years, there has been an increasing awareness that the enormous amount of data being captured by both public and private organisations can be profitably used for decision making. Aided by low-cost computer hardware, fast processing speeds and advancements in data storage technologies, Big Data Analytics has emerged as a fast growing field. However, the statistical challenges that are faced by statisticians and data scientists, while doing analytics with Big Data has not been adequately discussed. In this paper, we discuss the several statistical challenges that are encountered while analyzing Big data for management decision making. These challenges give statisticians significant opportunities for developing new statistical methods. Two methods—Symbolic Data Analysis and Approximate Stream Regression—which holds promise in addressing some of the challenges with Big Data are discussed briefly with real life examples. Two case studies of applications of analytics in management—one in marketing management and the other in human resource management—are discussed.

1 Introduction

If we look at the evolution of the adoption of analytics in businesses over time we see that there are several distinct stages. It all began in the early 1980s with 'standard reports' that described what had happened to a business function within a specified time period. These reports could be information about how many transactions took place in a branch, how much quantity of a product was sold, what was the amount spent on a certain stock keeping unit (SKU) by the loyalty card holders of a retailer, etc. The reports had to be preprogrammed and were produced at specified times such as quarterly, half-yearly or annually. With the advent of relational database systems (RDBMS) and the structured query language (SQL) ad-hoc reporting became possible. The managers could query their organisation's databases to seek information and reports which were not preprogrammed and were not routinely required. The

A. Laha (✉)
Indian Institute of Management Ahmedabad, Ahmedabad, India
e-mail: arnab@iima.ac.in

© Springer India 2016
S. Pyne et al. (eds.), *Big Data Analytics*, DOI 10.1007/978-81-322-3628-3_3

ability to query the database and drill-down as and when required allowed managers to figure out root causes of problems and helped them in eliminating these. It also enabled creation of alerts that were triggered to draw attention of managers for taking appropriate action when necessary. To give an example of such an alerting system, think of a bank customer approaching a teller with a request for encashing a cheque. The system would automatically alert the concerned bank manager for clearance of the cheque by sending a message on his desktop. Once the manager cleared the check the system would then alert the teller that the approval has been obtained through a message on the teller's desktop who can then make the payment to the customer.

In this context, it may be noted that much of the development of statistics in the early years of the twentieth century was driven by applied problems coming from different areas such as manufacturing, agriculture, etc. Two prominent examples being the work of Walter Shewhart who developed statistical process control (SPC) at the Western Electric's Hawthorne plant and the work of Ronald A. Fisher, who in the context of agricultural field experiments at Rothamstead Experimental Station, developed the field of design of experiments (DoE). Both of these techniques continue to be used by industries across the world for improving their quality and productivity. Seeing the great potential of using statistics in improving the quality of life in an emerging country like India, Mahalonobis [13] described statistics as a key technology and advocated its wider adoption in formulation of public policy and in solving challenging problems facing the country.

In the later years of the 1990s, statistical models started to be used much more widely than before in organisations to help them gain competitive advantage over their rival organisations. Since the early adopters reported substantial gains because of the use of analytical models, adoption of analytics for business decision making accelerated and spread to a variety of industries such as consumer products, industrial products, telecommunications, financial services, pharmaceuticals, transport, hospitality and entertainment, retail, e-commerce, etc. ([6] p. 7). These statistical models primarily helped managers by giving them insights (why a certain phenomena happened and what would happen if some things were done differently), forecasting (demand for different SKUs, sales, profit, etc.) and predicting occurrence or non-occurrence of some events (like attrition of customers, defaults on loans, etc.). Some organisations like American Airlines could use the vast data available with them to increase their revenue by adopting practices such as dynamic pricing and overbooking using statistical demand forecasting and optimization techniques [23]. Hewlett-Packard could ensure that on its online sales portal HPDirect.com, the right customer received the right offer at the right time leading to higher conversion rates and order sizes [26].

The use of statistical techniques on large databases for business decision making is now a common phenomenon in the developed world and is fast catching up in emerging economies. The data gathering abilities has increased tremendously over the last decade. With data coming from a variety of sources for example posts to social media sites like Facebook or Twitter, digital pictures and video uploaded by users, streaming video from surveillance cameras, sensors attached to different devices, mobile phone records, purchase transaction records, searches conducted on websites such as

Google the volume of data potentially available for decision making is truly huge. It is commonly estimated that more than 90 % of the total data generated by mankind has been generated in the last two years. It is therefore not surprising that the statistical methods developed in an era where even a sample of size 30 was considered large would face challenges when applied to extract information from such very large databases. As we shall see later in this article volume of data is not the only challenge that is faced by statisticians dealing with 'Big Data'. In fact with generous support coming from scientists working in the field of computer science, computational challenges can hopefully be overcome much sooner than the other challenges. Since with every challenge comes an opportunity, 'Big Data' offers statisticians opportunities for developing new methods appropriate for dealing with these kind of data.

2 Big Data

One of the key characteristics of Big Data used for business decision making is that it is often collected for a purpose, which is different from the purpose for which it is being analysed. In fact, Landefield ([11], p. 3) defines Big Data as the 'use of large-scale business and administrative data sets that are being used for secondary purposes other than those for which the data was ordinarily collected'. It is this aspect of Big Data that makes proper statistical analysis and decision making using Big Data particularly challenging.

Big Data is usually characterised by 4Vs—Volume, Velocity, Variety and Veracity. The total Volume of data is generally very high—several terabytes or more, the data acquisition rate (Velocity) is generally very high, different types of data (such as numerical, categorical, ordinal, directional, functional, text, audio, images, spatial locations, graphs) may be present in the same database (Variety) and these different types of data may be related with one another, and finally poor quality (Veracity) of the data because it is commonly seen that the data has been collected over a period of time by different agencies for different purposes. The final V is particularly challenging as it is difficult to clean such high volumes of data with more data coming in from multiple sources.

The key enablers in the present growth of Big Data are rapid increase in storage capabilities making storage inexpensive, rapid increase in processing power and developments in distributed storage mechanisms, availability of huge volume of data from diverse sources, and last but not least, a better value for money proposition for businesses. To understand why Big Data is a better value for money proposition for businesses, let us consider a few examples below.

Everyday, a large volume of messages are uploaded on social networking sites such as Facebook, Twitter, etc. These messages provide a storehouse of opinions and sentiments about a variety of products, services, regulatory policies, effectiveness of governance, etc. Sentiment analysis aims at analysing such vast amount of messages using techniques of text analytics to find out the opinion of the target customer segment of a product or service. These opinions or sentiments can be classified as positive, negative or neutral and managerial action can be initiated on this basis. A change in sentiment of the customers in the target segment of a product (or service) from positive to negative needs to be immediately addressed to because it presents an opportunity to competitors to wean away the customers. Hence, most large companies operating in consumer durables or fast moving consumer goods (FMCG) industry actively monitor the sentiment of their customers and aim to maintain a high positive rating. You may note that all the 4Vs pose a challenge to analysis of such data.

Another example of use of analytics is in predicting customer churn in the highly competitive telecom industry. Since customer acquisition cost in this industry is quite large companies aim to retain their most valuable customers through a variety of means such as giving attractive service offers, discounts, freebies, etc. For almost all companies in this industry, a large number of their customers are occasional users of their services and do not generate high revenues whereas only a few of them generate the bulk of their revenues. Early identification of churn intentions of any member of this group of high revenue generators helps the company in taking appropriate steps to prevent the churn. Analysis of the very large volume of call records often throws up interesting patterns which enables these companies to provide customised offers of value to these customers thereby retaining them. Any social media message carrying negative sentiment about the service or positive sentiment about a competitor's service from a member of this select customer group can often be seen as a leading indicator about their churn intention. Thus, one needs to combine information of one's own database, here for example the call record details, with the information about the same customer's behaviour on the social networking sites, to develop a more holistic view of the customer's level of satisfaction with the service and gauge the extent of his churn intention. Again, in this case also all the 4Vs pose challenge to analysis of this data.

As a third example, consider the problem of stopping credit card frauds. Analytics play an important role in inferring from the customers past behaviour whether the present transaction is genuine or is a possible case of fraud which needs to be stopped. Using transaction related information available with the credit card company such as the amount, the place where the card is swiped, time, merchant history, etc., and comparing the same with the customer's past transaction records a probability of fraud for the current transaction can be calculated. If this probability turns out to be large the transaction is denied and a message is sent to the customer asking him/ her the validity of the attempted transaction or seeking identification details. If the sought information is successfully provided then the transaction is subsequently allowed else a note is taken about this fraud attempt and the company may replace the credit card of the customer immediately with a new card. The entire exercise of determining whether an alert should be raised or not has to be done within a few seconds making this apparently straightforward binary classification task challenging in the context of Big Data.

As the fourth and final example, let us consider the huge stream of videos that are available from the surveillance cameras of a large facility like an airport. The aim of surveillance at the airport is mainly to prevent happening of untoward incidents by identifying suspicious movements, unclaimed objects and presence of persons with known criminal records. The task here is humongous akin to finding a needle in a haystack. Video analytics can be used to analyse video footages in real time for detecting the presence of suspicious persons or unclaimed objects amongst the thousands of genuine travellers. In this case the volume, velocity and variety of data pose a challenge to real-time analysis of this data.

3 Statistical Challenges

In this section, we discuss different challenges that are encountered while analyzing big data from a statistical perspective. We structure our discussion around the 4Vs mentioned in the earlier section.

3.1 Volume

As we have discussed earlier, advancement in data acquisition and storage technology has made it possible to capture and store very large amounts of data cheaply. However, even with very high speed processors available today some of the simple statistical methods which involve sorting (such as the computations of quantiles) or matrix multiplication (such as linear regression modelling) face computational challenges when dealing with Big Data. This is so because the popular algorithms for sorting or matrix multiplication do not scale-up well and run very slow on terabyte-scale data sets. For example, the very popular sorting algorithm, *Bubble Sort* is not suitable for use with large data sets since its average and worst case time complexity are both $O(n^2)$ where n is the number of data elements. Instead, it would be better to use an algorithm such as *Heapsort* whose average and worst case time complexity are both $O(n \log n)$ or the *Radix Sort* whose average and worst case time complexity are both $O(nk)$ where k in general is a function of n [5] (see Cormen et al. 2009 for more details regarding sorting algorithms and their computational complexity).

Matrix multiplication is ubiquitous in statistics right from the theory of linear models to modern advanced methods such as Nonparametric Regression. However, the computational complexity of matrix multiplication is high since for computing $C_{m \times n} = A_{m \times k} B_{k \times n}$ using the standard method we need $O(mkn)$ computations. In the special case when $m = k = n$, this reduces to $O(n^3)$ computations. Thus in the full-rank Gauss–Markov model $Y_{n \times 1} = X_{n \times k} \beta_{k \times 1} + \epsilon_{n \times 1}$ the computation of the *BLUE* of β which is $\hat{\beta}_{k \times 1} = (X^T X)^{-1} X^T Y$ becomes very time consuming for large n (see Rao, 1973 for details of the Gauss–Markov model and its applications). Hence, faster algorithms such as the Strassen's algorithm [24] or the Coppersmith–Winograd algorithm [4] need to be used for carrying out such computations with very large data.

3.2 Velocity

Some of the fundamental ideas of statistics come under challenge when dealing with Big Data. While 'optimality of statistical procedures' has been the ruling paradigm in Statistics for decades, the speed of computation has never been of primary concern to statisticians while developing statistical methods until very recently. In management applications, it is often expected that the acquired data would be analysed in the shortest possible time so that necessary action can be taken as quickly as possible. For example, in real-time surveillance for financial fraud, it is important to conclude whether a proposed transaction is a fraudulent one or not in a matter of few seconds. Analysis of such 'streaming data' opens up new challenges for statisticians as it is not possible to come to a decision after analyzing the entire dataset because of severe time constraints. A general feature of analysis of streaming data is that a decision regarding whether there is some change in the data generating mechanism is sought to be arrived at by looking only at a very small part of the available data. This calls for a paradigm shift regarding the choice of statistical procedures when dealing with streaming data. Instead of choosing statistical procedures based on the present paradigm of 'Get the best answer' we need to choose procedures based on the paradigm 'Get a good answer fast'.

Streaming data is not entirely new to the statistical community. In online process control, it is assumed that the data is streaming but possibly at a much slower rate than in the Big Data applications. The widely used statistical process control (SPC) tools such as $\bar{X}$-chart, CUSUM chart, EWMA charts, etc., have been devised keeping in mind the streaming nature of data (see [16] for details regarding control charts). Thus, these techniques with suitable adaptations and modifications have potential to be applied with streaming data even in the Big Data context. One of the key differences of the streaming data in the Big Data context with that in the manufacturing context is that in the latter context a probabilistic structure for the streaming data can be assumed which may not even exist in the former context and even if it does it may be highly non-stationary [27]. While designing algorithms for working with data streams, it is important to remember that data cannot be permanently stored. This implies that the algorithms will have to discard the data after using them. Another requirement is that the results should be swiftly updated with current data getting more weight than historical data. To understand this point, let us suppose our interest is in getting an estimate of the current mean of the data stream. The computation of arithmetic mean of a data stream can be done simply by recursively computing

$$\bar{X}_n = \frac{n-1}{n}\bar{X}_{n-1} + \frac{X_n}{n}$$

but this responds very slowly to any change in the process mean as it gives very low weight to the current data compared to historical data. An alternative is to use the exponentially weighted moving average (EWMA) Y_n which can be computed recursively as

$$Y_n = \theta Y_{n-1} + (1 - \theta)X_n, 0 < \theta < 1.$$

It is easy to see that Y_n responds to any change in mean of the data stream much faster than $\bar{X}_n$, while giving the correct mean when there is no change.

Often, Big Data does not come from a single population but is a mixture of data coming from many populations. This problem is not entirely new and procedures for dealing with mixture of populations, such as cluster analysis, has been in existence for quite some time. Standard statistical procedures such as linear regression analysis does not perform well when dealing with data which is a mixture of different populations. A naive approach is to first cluster the data using a clustering algorithm such as K-means algorithm and then carry out linear regression in each of these clusters separately treating each cluster as a 'population'. Desarbo and Cron [7] suggested a method for doing clusterwise linear regression using a maximum likelihood approach using the mixture of Gaussian distributions setup. In this case the regression parameters in each cluster are simultaneously estimated. Alternative nonparametric approaches include procedures such as classification and regression trees popularly known as CART [3].

The additional challenge that streaming data poses is that the number of populations (clusters) may not be known in advance and that it may vary over time. This is often seen in social media data where new groups with distinct characteristics suddenly emerge and sometimes established groups decay and eventually disappear. Thus finding estimates, trends or patterns can become a very difficult exercise introducing unique computational and statistical challenges. Silva et al. [22] gives a comprehensive survey of the work done in the area of clustering data streams data in recent years.

3.3 Variety

Big data is often a mixture of different types of data leading to the problem of Variety. The well-known statistical methods either deal with numerical or categorical data. Thus when dealing with mixture of data types such as video data and categorical data, the current statistical methods are not of much use. To give an example of how such data may arise in practise consider the following situation: A video is requested by an user from a website such as Youtube and suppose an advertisement video precedes the requested video. The user may watch the advertisement video in full or watch it partially or may entirely skip it. A creative content company may be interested in knowing the characteristics of the advertisement video that would induce the user to see it in full or with what kind of videos the advertisement video should be bundled so that the chance the user sees it to the full is maximised. Situations where multiple types of data occur together are also quite common in many areas of science such as speed and direction of wind in meteorology (linear data and angular data), variation in shape of skull across humans of different ethnicity (shape data and categorical data), variation of blood pressure and development of heart disease (functional data and categorical data), etc. Over the last two decades, our knowledge of analyzing non-numerical non-categorical data has increased substantially. Mardia and Jupp [15] give a book length treatment of analysis of directional data with particular emphasis on analyzing circular and spherical data (see also [9]). A detailed account of the analysis of shape data can be found in Dryden and Mardia [8]. Bhattacharya and Bhattacharya [1] discusses nonparametric inference on manifolds. Ramsey and Silverman [19] give an excellent introduction to functional data analysis. More such new methodological developments are needed for drawing sound inference with data of different types. The field of Symbolic Data Analysis holds much promise in addressing the Variety problem encountered with Big Data as it deals with different data types together (see Billard [2] and Noirhomme-Fraiture and Brito [18] for more details).

3.4 Veracity

Veracity of data is a big concern when dealing with Big Data. The data can be inaccurate because of many possible reasons including intentional falsification. In reviews of hotels, films, cars, etc. it is difficult to decide whether the review has been given in a fair manner or is motivated by reasons other than the merit of the product or service. Though it is difficult to weed out false reviews completely many websites are attempting to develop checks which identifies and weeds out false reviews. Also mechanisms are being evolved to deter people from writing false reviews. However, at present there are hardly any statistical methods available that can detect and weed out false data from a given data set automatically. Thus, it is presently a herculean task to clean up such large volumes of data before doing any analysis. Better automated methods for identifying and eliminating incorrect data are an urgent requirement.

Karr et al. [10] defines Data Quality as 'the capability of data to be used effectively, economically and rapidly to inform and evaluate decisions'. Since the Big Data are often collected for a purpose different from that for which it is being analysed, there is no sampling design and it is well known to statisticians that such data may induce bias which may lead to conclusions which are not generalizable. Moreover, there is possibility that the data definitions as well as the mechanism of data collection may have changed over time. For example in credit default prediction studies the persons seeking loan from an institution may change over a period of time based on bank's behaviour in terms of giving loans. Hence, Data Quality is often poor in Big Data.

Most of the currently known statistical methods rely on the assumption that the data being analysed is a random sample from the underlying population. This assumption is generally not valid when working with Big Data in the field of Management. In almost all encounters with Big Data in Management it is seen that the data is non-random. Therefore, it is of great importance to validate any insight that is obtained through analysis of Big Data, using data which has been collected using a statistically sound method before taking any important decision based on that insight.

3.5 Privacy and Confidentiality

Last but not the least is the concern that Big Data analytics can threaten privacy and confidentiality and can impinge on an individual's personal freedom. In recent years we have seen several instances where an organisation or a state agency has attempted to predict the status of an individual for reasons ranging from purely commercial to involving national security using Big Data. Such use of Big Data raises substantial questions of ethics and calls for new frameworks and guidelines to prevent misuse of such data.

In many situations, particularly when organisations have to release data for public use, it becomes important to treat the data in a manner that prevents recovery of individual level confidential information from the released data. Reiter [20] discusses in detail the statistical approaches to protecting data confidentiality that are commonly used and the impact that it has on the accuracy of secondary data analysis. Since most data that are currently generated is being stored with the expectation that its analysis in future, in possibly different context, may help in generating new insights for the organisation, it is extremely important that such data be treated for protecting data confidentiality before storage.

4   Statistical Methods for Big Data

In the earlier section, we have discussed the various statistical challenges that are brought forth by Big Data. In this section, we briefly discuss two statistical methods namely Symbolic Data Analysis and Approximate Stream Regression which can be used to tackle the Variety and Velocity problems of Big Data, respectively, in some applications.

4.1 Symbolic Data Analysis

Symbolic data arise in Big Data context through aggregation of data for certain features of interest. For example, a credit card company need not store the entire history of credit card transactions of its customers but instead may store only the minimum and maximum amounts transacted by them. Thus for each customer the bank stores an interval, which is the range of the amounts transacted by the customer. Such data are refereed to as *interval-valued* symbolic data. Another possibility is that it creates a histogram (or frequency distribution) of the transaction amounts and stores the same. In this case the data is *histogram-valued*. There can be many other forms of symbolic data as discussed in Billard [2]. In the following we restrict ourselves to interval-valued symbolic data. Let $[a_i, b_i], i = 1, \ldots, n$ be a random sample of n intervals. Assuming that the variable of interest (say, credit card transaction amounts) of the i-th customer are uniformly distributed in the interval $[a_i, b_i]$ we can compute the mean $\Theta_{1i} = \frac{a_i+b_i}{2}$ and the variance $\Theta_{2i} = \frac{(b_i-a_i)^2}{12}$. The random variables $\Theta_i = (\Theta_{1i}, \Theta_{2i})$ are referred to as the internal parameters of the i-th interval $[a_i, b_i]$. Le-Rademacher and Billard [12] assumes that the random variables Θ_{1i} and Θ_{2i} are independent with Θ_{1i} distributed as $N(\mu, \sigma^2)$ and Θ_{2i} distributed as $Exp(\beta)$ for all $i = 1, \ldots, n$. They obtain the MLEs of μ, σ and β based on the observed interval-valued data. As an illustration, we apply this technique to examine the temperature of India in the months of October–December. Historical data of the maximum and minimum temperatures during these months for the years 1901–2011 are available from the weblink

```
https://data.gov.in/catalogs/ministry_department/india-
meteorological-department-imd
```

We compute the internal parameters Θ_{1i} and Θ_{2i} for the given data using the formula discussed above. We find that it is reasonable to assume both Θ_{1i} and Θ_{2i} to be normally distributed using the *Ryan-Joiner* normality test implemented in *Minitab 16* software. We assume Θ_{1i} to be distributed as $N(\mu, \sigma^2)$ and Θ_{2i} to be distributed as $N(\nu, \tau^2)$. The MLEs of the parameters are found to be $\hat{\mu} = 21.82, \hat{\sigma} = 0.44, \hat{\nu} = 9.44$ and $\hat{\tau} = 0.91$. Now note that if X is distributed as $Uniform(a, b)$ then $a = E(X) - \sqrt{3}SD(X)$ and $b = E(X) + \sqrt{3}SD(X)$. Hence, we may consider the interval $[21.82 - \sqrt{3} \times 9.44, 21.82 + \sqrt{3} \times 9.44] = [16.50, 27.14]$ as the average range of temperature during the months of October–December in India.

4.2 Approximate Stream Regression

In many applications such as predicting the amount of monthly spend by a customer on a certain category of products, multiple regression models are used. As the data accumulates over time, often a phenomenon referred to as *concept drift* is encountered. In simple words, this means that the regression coefficients may change over a period of time. This shows up in larger prediction errors leading to rebuilding of the regression model with current data. However, it is preferable to have a method which adapts to the concept drift, i.e. the regression coefficients are automatically updated to its current values without needing to rebuild the model which is often a costly exercise. Nadungodage et al. [17] proposes the Approximate Stream Regression (ASR) to address this problem. One of the key concerns with high velocity streaming data is that the entire data cannot be stored. Thus the data which has been used has to be discarded and cannot be used again in future. In ASR the data is accumulated in blocks, say of size m. Let β_k^* be the vector of regression coefficients computed using the data of the k-th block only and let β_{k-1} be the vector of regression coefficients computed at the end of processing of the first $k-1$ blocks, then β_k is computed according to the formula,

$$\beta_k = (1 - \alpha)\beta_k^* + \alpha\beta_{k-1}$$

where $0 < \alpha < 1$. It is easy to see that

$$\beta_k = (1 - \alpha)\beta_k^* + \alpha(1 - \alpha)\beta_{k-1}^* + \alpha^2(1 - \alpha)\beta_{k-2}^* + \ldots + \alpha^{k-2}(1 - \alpha)\beta_2^* + \alpha^{k-1}\beta_1^*.$$

The value of α determines the emphasis that is given on current data. A value of α close to 0 indicates that the current data is emphasised more, whereas a value of α near 1 indicates that the historical data is given more importance. We now illustrate using an example the advantage of using ASR over the common approach of building a regression model and then using it for prediction in the context of streaming data with concept drift. For this purpose we generate six blocks of data with each block having 30 observations on Y, X_1 and X_2. The variable Y is taken as the response variable and the variables X_1 and X_2 are taken as predictor variables. The variables X_1 and X_2 are generated from normal distribution with mean 0 and standard deviation 1. In the first three blocks we compute Y as $Y = X_1 + X_2 + \epsilon$, whereas in the remaining three blocks we compute Y as $Y = 1 + 2X_1 - X_2 + \epsilon$ where the errors (ϵ) are generated from normal distribution with mean 0 and standard deviation 0.5. All computations are done using *Minitab 16* software. The value of α for ASR is fixed to be 0.75. The ASR method is compared with the linear regression model built by using only the data for Block 1. For this method the predicted values in Blocks 2–6 are obtained using the regression coefficients estimated on the basis of Block 1 data. The two methods are compared on the basis on Mean Square Predictive Error (MSPE). The estimated regression coefficients in each block, the regression coefficients obtained using the ASR method and the MSPE of the ASR method and that of the fixed coefficients method are given in Table 1.

Table 1 Comparison of predictive performance of ASR with linear regression for streaming data with concept drift ($\alpha = 0.75$)

Block	$\hat{\beta}_0$	$\hat{\beta}_1$	$\hat{\beta}_2$	MSPE ASR	MSPE Fixed	$\hat{\beta}_0$ ASR	$\hat{\beta}_1$ ASR	$\hat{\beta}_2$ ASR
1	0.095	0.932	0.829					
2	−0.054	0.965	1.02	0.362	0.362	0.058	0.94	0.877
3	0.046	1.05	0.992	0.239	0.262	0.055	0.968	0.906
4	1.08	1.94	−0.965	7.13	6.74	0.311	1.211	0.438
5	1.09	1.96	−0.905	1.87	2.967	0.506	1.398	0.102
6	0.99	2.1	−1.04	1.68	4.37	0.627	1.574	−0.183

It is seen that the ASR method performs much better than the fixed coefficients method in terms of having lower MSPE after the information of the changed regression coefficients are fed into the ASR model. Further it is seen that the regression coefficients of the ASR method steadily move towards the new values as more information become available. Thus it is expected that ASR would prove to be a useful method for doing regression with streaming data.

5 Case Studies

In this section, we look at two case studies of application of Big Data in management. The first case study looks at the problem of cross selling in an online environment without any human intervention and suggest a methodology for the same and the second case study discusses the improvements achieved in the talent acquisition process of a large Indian IT company using innovative analytical solutions.

5.1 Online Recommendations for Cross Selling

In today's competitive environment, cost of acquiring a customer is ever rising. Hence, organisations are attempting to realise better value from their existing customers by selling other useful products (or services) to them. For this purpose, it is important for an organisation to identify the products (or services) that would be most beneficial to each individual customer and to provide him/her with the best value proposition for the same. Many organisations use analytics to find out the best offer for each customer and then use human customer service representatives to communicate the same to him/her. However this is a costly process and with the advent of 'Do not Call' registries in countries such as India, it now faces some regulatory challenges as well. Moreover, a significant number of customers now prefer to use the websites of the organisations to seek information and carry out transactions. Thus, it is necessary to have automatic recommendation systems that would aid in cross selling in an online environment. Majumdar and Mukherjee [14] describe the use of online cluster analysis in determining the most appropriate offers to be given to a customer based on the available information. They convert the characteristics of each customer into a vector. For each customer visiting the website, they compute the *cosine distance* of this customer with all the customers in their training dataset based on the information about the customer available with them. The *cosine distance* between two customers having vector representation $\tilde{x}$ and $\tilde{y}$ is given by $\frac{<\tilde{x},\tilde{y}>}{|\tilde{x}||\tilde{y}|}$ where $< \tilde{x}, \tilde{y} >$ is the dot product between the two vectors $\tilde{x}$ and $\tilde{y}$ and the $|.|$ denotes the length of a vector. Note that the cosine distance takes the maximum value of $+1$ when $\tilde{x} = \tilde{y}$ and hence is not a distance in the usual sense. In what follows I use the term *cosine similarity* instead of cosine distance. The products purchased by the 'micro-cluster' of customers in the training dataset having the maximum (or close to the maximum) cosine similarity with the current customer are analysed. Each product is broken down in terms of their attributes and each product attribute is given a weight. Suppose there are m product attributes and n customers in the micro-cluster. A simple weighting scheme is to give the attribute A_j the weight w_j, where

$$w_j = \frac{\sum_{i=1}^{n} f_{ij} d(i)}{\sum_{j=1}^{m} \sum_{i=1}^{n} f_{ij} d(i)}.$$

Here f_{ij} is the frequency of occurrence of attribute A_j among the purchases of the ith customer in the micro-cluster and $d(i)$ is the cosine similarity of the ith customer with the current customer. The current customer is then recommended products that have the attributes that have been assigned the highest weights. One of the problems that would be faced by this method would be that it would require frequent updating of the training dataset as the customers' buying preferences is likely to change over time.

5.2 Talent Acquisition

Large service organisations in the IT-sector in India need to recruit very large number of people every year to maintain their growth and account for attrition. Tata Consultancy Services (TCS) which is a large company having more than 300,000 lakh employees recruited approximately 70000 new staff in the fiscal year 2012–2013. This requires huge investment in attracting the right kind of people to apply, screening these applications, interviewing them, selecting the most deserving of them, create offers which attract the selected candidates to join the organisation and finally train them and make them ready for their job. Srivastava [25] discusses the use of analytics for improving the talent acquisition process at TCS with the goal of getting the required number of good quality new hires at a reduced overall cost meeting the lead time specification for each role. Working on a dataset of 26574 candidates who were interviewed by TCS-BPO during the year 2010 they use decision trees to predict (i) whether a candidate will be selected if interviewed and (ii) if a selected candidate will join the organisation. They use linear regression modelling to predict the time of joining of a selected candidate and also carry out 'subgroup discovery' to identify the characteristics of candidates who take much longer time to join compared to the others. Further, since offer declines are very costly a Bayesian net is used to find out the root cause of the offer declines, so that by addressing these causes offer declines could be reduced.

It is not difficult to see that the talent acquisition process described above can be further improved if unstructured data such as audio/video recordings of interviews, statements about projects worked on, past experience, career objectives, etc., made in the candidate's resume (or application form) could be used for model development.

However, better methods of extracting information from unstructured data coming from multiple sources would be required to make this possible.

6 Concluding Remarks

In this paper, we have discussed some of the statistical challenges that are faced by statisticians and data scientists while doing analytics with Big Data. These challenges give statisticians significant opportunities for developing new statistical methods that (a) are scalable and can handle large volumes of data easily, (b) analyse streaming data and give *'Good answers fast'*, (c) handle multiple data types and help in understanding association amongst data of different types, (d) can derive meaningful results even with poor quality data and (e) protect privacy and confidentiality of individual level data without impacting the accuracy of aggregate level analysis. We have also briefly discussed two methods Symbolic Data Analysis and Approximate Stream Regression, which holds promise in addressing some of the challenges with Big Data. In addition, two case studies of applications of analytics in management—one in marketing management and the other in human resource management—are discussed.

References

1. Bhattacharya A, Bhattacharya R (2012) Nonparametric inference on manifolds: with applications to shape spaces. Cambridge University Press, Cambridge
2. Billard L (2011) Brief overview of symbolic data and analytic issues. Stat Anal Data Min 4(2):149–156
3. Breiman L, Friedman JH, Olshen RA, Stone CJ (1984) Classification and regression trees. Wadsworth Inc
4. Coppersmith D, Winograd S (1990) Matrix multiplication via arithmetic progressions. J Symb Comput 9:251–280
5. Cormen TH, Leiserson CE, Rivest RL, Stein C (2009) Introduction to algorithms, 3rd edn. MIT Press and McGraw Hill
6. Davenport TH, Harris JG (2007) Competing on analytics: the new science of winning. Harvard Business School Publishing Corporation
7. Desarbo WS, Cron WL (1988) A maximum likelihood methodology for clusterwise linear regression. J Classif 5:249–282
8. Dryden IL, Mardia KV (1998) Statistical shape analysis. Wiley
9. Jammalamadaka SR, Sengupta A (2001) Topics in circular statistics. World Scientific
10. Karr AF, Sanil AP, Banks DL (2006) Data quality: a statistical perspective. Stat Methodol 3(2):137173
11. Landefield S (2014) Uses of big data for official statistics: privacy, incentives, statistical challenges, and other issues. Discussion Paper, International conference on big data for official statistics, Beijing, China, 28–30 Oct 2014. http://unstats.un.org/unsd/trade/events/2014/beijing/SteveLandefeld-UsesofBigDataforofficialstatistics.pdf. Accessed 30 May 2015
12. Le-Rademacher J, Billard L (2011) Likelihood functions and some maximum likelihood estimators for symbolic data. J Stat Plan Inference 141:1593–1602

13. Mahalonobis PC (1965) Statistics as a key technology. Am Stat 19(2):43–46
14. Majumdar K, Mukherjee S (2011) Designing intelligent recommendations for cross selling. In: Video documentation of 2nd IIMA International conference on advanced data analysis, business analytics and intelligence, DVD-II, IIM Ahmedabad, India
15. Mardia KV, Jupp PE (1999) Directional statistics. Wiley
16. Montgomery DC (2012) Statistical quality control, 7th edn. Wiley
17. Nadungodage CH, Xia Y, Li F, Lee JJ, Ge J (2011) StreamFitter: a real time linear regression analysis system for continuous data streams. In: Xu J, Kim MH, Unland R (eds) Database systems for advanced applications, pp 458–461. Springer
18. Noirhomme-Fraiture M, Brito P (2011) Far beyond the classical models: symbolic data analysis. Stat Anal Data Min 4(2):157–170
19. Ramsey JO, Silverman BW (2005) Functional data analysis, 2nd edn. Springer
20. Reiter JP (2012) Statistical approaches to protecting confidentiality for microdata and their effects on the quality of statistical inferences. Public Opin Q 76(1):163–181
21. Rao CR (1973) Linear statistical inference and its applications, 2nd edn. Wiley
22. Silva JA, Faria ER, Barros RC, Hruschka ER, de Carvalho ACPLF, Gama J (2013) Data stream clustering: a survey. ACM Comput Surv 46, 1, 13:1–13:31
23. Smith BC, Leimkuhler JF, Darrow RM (1992) Yield management at American Airlines. Interfaces 22(2):8–31
24. Strassen V (1969) Gaussian elimination is not optimal. Numerische Mathematik 13:354–356
25. Srivastava R (2015) Analytics for improving talent acquisition process. In: Video documentation of 4th IIMA International conference on advanced data analysis, business analytics and intelligence, DVD-II, IIM Ahmedabad, India
26. Tandon R, Chakraborty A, Srinivasan G, Shroff M, Abdullah A, Shamasundar B, Sinha R, Subramanian S, Hill D, Dhore P (2013) Hewlett Packard: delivering profitable growth for HPDirect.com using operations research. Interfaces 43(1):48–61
27. Wegman EJ, Solka JL (2005) Statistical data mining. In: Rao CR, Wegman EJ, Solka JL (eds) Data mining and data visualization, handbook of statistics, vol 24. Elsevier

Application of Mixture Models to Large Datasets

Sharon X. Lee, Geoffrey McLachlan and Saumyadipta Pyne

Abstract Mixture distributions are commonly being applied for modelling and for discriminant and cluster analyses in a wide variety of situations. We first consider normal and t-mixture models. As they are highly parameterized, we review methods to enable them to be fitted to large datasets involving many observations and variables. Attention is then given to extensions of these mixture models to mixtures with skew normal and skew t-distributions for the segmentation of data into clusters of non-elliptical shape. The focus is then on the latter models in conjunction with the JCM (joint clustering and matching) procedure for an automated approach to the clustering of cells in a sample in flow cytometry where a large number of cells and their associated markers have been measured. For a class of multiple samples, we consider the use of JCM for matching the sample-specific clusters across the samples in the class and for improving the clustering of each individual sample. The supervised classification of a sample is also considered in the case where there are different classes of samples corresponding, for example, to different outcomes or treatment strategies for patients undergoing medical screening or treatment.

Keywords Flow cytometry · Sample of cells · Multiple samples · Clustering of cells · Supervised classification of samples · Skew mixture models · EM algorithm

1 Introduction

Finite mixture distributions are being increasingly used to model observations on random phenomena in a wide variety of applications. Many of these applications use normal mixture models, which can be fitted easily at least for independent data using the expectation–maximization (EM) algorithm of Dempster et al. [1]. This

S.X. Lee · G. McLachlan (✉)
Department of Mathematics, University of Queensland, St. Lucia, QLD 4072, Australia
e-mail: g.mclachlan@uq.edu.au

S. Pyne
Public Health Foundation of India, Indian Institute of Public Health, Hyderabad, India
e-mail: spyne@iiphh.org

© Springer India 2016
S. Pyne et al. (eds.), *Big Data Analytics*, DOI 10.1007/978-81-322-3628-3_4

seminal paper greatly simulated interest in the use of finite mixture distributions to model heterogeneous data. This is because the fitting of mixture models by maximum likelihood (ML) is a classic example of a problem that is simplified considerably by the EM's conceptual unification of ML estimation from data that can be viewed as being incomplete; see, for example, McLachlan and Peel [2].

One field in which the normal mixture model and its robust version using t-distributions as components have been applied is bioinformatics with particular attention to the analysis of microarray gene-expression data. This has involved problems such as the unsupervised classification (cluster analysis) of so-called tissue samples containing the expression levels of thousands of genes [3]. As the normal and t-mixture models are highly parametrized, consideration has to be given to ways in which they can still be fitted to big datasets in which the number of variables p is usually very large relative to the sample size n.

We shall give a brief account of the fitting of normal and t-mixtures to high-dimensional data, including the use of mixtures of factor models in which the component-covariance matrices are taken to have a factor-analytic representation. In some applications, the number of variables are too large for even mixtures of factor models to be fitted. Hence some form of dimension reduction has to be undertaken. We shall consider methods such as variable selection and projections of the data into a space of much lower dimension than p.

More recently, mixture models are being used in an attempt to provide an automated approach to the analysis of samples in flow cytometry. Flow cytometry is a powerful tool in clinical diagnosis of health disorders, in particular immunological disease. It offers rapid high-throughput measurements of multiple characteristics on every cell in a sample, capturing the fluorescence intensity of multiple fluorophore-conjugated reagents simultaneously. It is routinely used in clinical practices for the diagnosis of health disorders such as immunodeficiency diseases.

A critical component in the pipeline of flow cytometry data analysis is the identification of cell populations from a multidimensional dataset that consists of a very large number of cells (perhaps in the hundreds of thousands). Hence, there is the need to segment the cells into a number of clusters on the basis of markers measured on each cell. In the case of multiple samples of cells, there is the additional registration problem of matching the sample-specific clusters across the samples. In situations where there are different classes of samples, a further problem concerns the construction of a classifier for assigning an unclassified sample to one of the predefined classes. The focus is on the JCM (joint clustering and matching) procedure for carrying out these aforementioned tasks. This procedure was proposed by Pyne et al. [4] to enable the clustering and registration problems to be performed simultaneously rather than first clustering each sample and then having the problem of trying to match the sample-specific clusters as in the FLAME procedure proposed earlier by Pyne et al. [5]. With the JCM approach to registration for multiple samples and supervised classification in the case of a number of predefined classes of samples, a template is constructed for each class.

To establish some notation, we let Y denote a p-dimensional random vector consisting of p feature variables associated with the random phenomenon of interest.

We let also $y_1, \dots, y_n$ denote the n observed samples on Y. We shall refer to each y_j as a feature vector.

The $n \times p$ data matrix A can be expressed as

$$A = (y_1, \dots, y_n)^T. \tag{1}$$

For example, in the context of analyzing gene-expression data from n microarray experiments on p genes, the feature vector y_j contains the expression levels of the p genes resulting from the jth microarray experiment ($j = 1, \dots, n$). Usually, p is very large (in the thousands or tens of thousands) compared to n, which is often less than a hundred. The feature vector y_j is known in this context as the gene signature vector while the ith row of the data matrix A is referred to as the profile (vector) for the ith gene.

In the analysis of flow cytometry data, the feature vector y_j contains the measurements on p markers on the jth cell. In this application n is usually very large, for example, in the hundreds of thousands, while in the past the number of markers p would be small (say, less than five), which enabled manual gating to be undertaken. Currently available flow cytometers can measure up to about 20 markers simultaneously, and so there is a need for an automated approach. The next generation of mass cytometers are able to measure more than 40 markers.

2 Finite Mixture Models

With the mixture model-based approach to density estimation and clustering, the density of Y is modelled as a linear combination (or mixture) of g component densities $f_i(y)$ with unknown mixing proportions π_i ($i = 1 \dots, \pi_g$). That is, each data point y_j is taken to be a realization of the probability density function (p.d.f.) of the mixture model, given by

$$f(y_j) = \sum_{i=1}^{g} \pi_i f_i(y_j) \qquad (j = 1, \dots, n). \tag{2}$$

The mixing proportions must satisfy the conditions $\pi_h \geq 0$ and $\sum_{i=1}^{g} \pi_i = 1$, that is, they are nonnegative and sum to one.

In density estimation, the number of components g in the mixture density (2) can be taken to be sufficiently large to provide an arbitrarily accurate approximation of the underlying density function; see, for example, Qi and Barron [6]. For clustering purposes, each component in the mixture model (2) is usually taken to correspond to the distribution of the data in a single cluster. Then a probabilistic clustering of the data into g clusters can be obtained in terms of the fitted posterior probabilities of component membership for the data. The posterior probability that an observation with feature vector y_j belongs to the ith component of the mixture is given by

$$\tau_i(y_j) = \pi_i f_i(y_j) / f(y_j).$$

$$(3)$$

On specifying a parametric form $f_i(y_j; \theta_i)$ for each component density, the parametric mixture model

$$f(y_j; \Psi) = \sum_{i=1}^{g} \pi_i f_i(y_j; \theta_i)$$

$$(4)$$

can be fitted by maximum likelihood (ML). In the above, θ_i is the vector consisting the unknown parameters of the ith component of the mixture model. The vector $\Psi = (\xi^T, \pi_1, \dots, \pi_{g-1})^T$ contains the unknown parameters of the mixture model, where ξ consists of the elements of the θ_i known a priori to be distinct; the superscript T denotes vector transpose. We let Ω denote the parameter space for Ψ.

The maximum likelihood estimate (MLE) of Ψ, denoted by $\hat{\Psi}$, is given by an appropriate root of the likelihood equation. That is, $\hat{\Psi}$ is a solution for Ψ to the equation

$$\partial \log L(\Psi) / \partial \Psi = 0,$$

$$(5)$$

where $L(\Psi)$ denotes the likelihood function for Ψ, given by

$$L(\Psi) = \prod_{j=1}^{n} f(y_j; \Psi),$$

which is formed under the assumption of independence of the data. Solutions of the likelihood equation (5) corresponding to local maximizers of log likelihood function $\log L(\Psi)$ can be obtained via the expectation-maximization (EM) algorithm of Dempster et al. [1]; see also McLachlan and Krishnan [7].

In practice, the clusters in the case of Euclidean data are frequently assumed to be elliptical, so that it is reasonable to consider fitting mixtures of elliptically symmetric component densities. Within this class of component densities, the multivariate normal density is a convenient choice given its computational tractability. In this case, the ith component density $f_i(y_j; \theta_i)$ is given by

$$f_i(y_j; \theta_i) = \phi(y_j; \mu_i, \Sigma_i) \quad (i = 1, \dots, g),$$

$$(6)$$

where $\phi(y_j; \mu_i, \Sigma_i)$ denotes the p-dimensional multivariate normal distribution with mean vector μ_i and covariance matrix Σ_i. The normal mixture density is then specified as

$$f(y_j; \Psi) = \sum_{i=1}^{g} \pi_i \phi(y_j; \mu_i, \Sigma_i).$$

$$(7)$$

The normal mixture model (7) is sensitive to outliers in the data since it adopts the multivariate normal family to model the distributions of the errors. To improve the robustness of this model for the analysis of data, which have longer tails than the normal or contain atypical observations, an obvious way is to consider using a heavy-tailed distribution such as the multivariate t-distribution. The t-distribution, along with some other well-known distributions such as the Cauchy distribution and Pearson type VII distribution, can be obtained by embedding the normal distribution in a wider class of elliptically symmetric distributions. With the t-distribution, there is an additional parameter v_i known as the degrees of freedom [8]. This parameter v_i may be viewed as a robustness tuning parameter that regulates the thickness of the tails of the distribution. When fitting the t-mixture model, this parameter can be fixed in advance (that is, known a priori) or it can be inferred from the data as part of the model fitting process.

One attractive feature of using mixture models that adopt elliptically symmetric distributions as components densities (such as the aforementioned normal and t-mixture models) is that the implied clustering is invariant under affine transformations. Also, in the case where the components of the mixture correspond to externally defined subpopulations, the unknown parameter vector Ψ can be estimated consistently by a sequence of roots of the likelihood equation. Concerning the issue of how to select an appropriate number of components for a mixture model, this can be approached by consideration of the value of the log likelihood function. For example, one may consider performing model selection using criteria such as the Bayesian Information Criterion (BIC) of Schwarz [9]. Another method is to employ a resampling approach such as [10] to test for the smallest value of g compatible with the data.

3 Factor Models

For datasets where the dimension p of the observation or feature vector is large relative to the number of observations n, the normal mixture model (7) may not be able to be fitted directly to the observed data. This is due to the normal mixture model (7) with general component-covariance matrices being a highly parameterized model. With unrestricted covariance matrices, there are $\frac{1}{2}p(p + 1)$ distinct elements in each Σ_i. A simple way of proceeding in such situations would be to take the Σ_i to be diagonal. However, this leads to the restriction that the axes associated with the clusters must be aligned with those of the feature space, which may not be appropriate in practice as the clusters are typically of arbitrary orientation. Further restricting the Σ_i to be a multiple of the identity matrix leads to spherical clusters. If the mixing proportions are taken to be the same, then it leads to a soft-version of k-means.

A common approach to reducing the dimensionality of the data is to perform a principal component analysis (PCA). The latter, however, provides only a global linear model for the representation of the data in a low-dimensional subspace. A more general and powerful approach is to adopt a mixture of factor anlayzers (MFA), which is a global nonlinear approach obtained by postulating a finite mixture of linear (factor) submodels for the distribution of the full observation vector y_j given a relatively small number of (unobservable) factors (McLachlan and Peel [11], Chap. 8). That is, a local dimensionality reduction method is provided by imposing the factor-analytic representation

$$\Sigma_i = B_i B_i^T + D_i. \tag{8}$$

In the above, B_i is a $p \times q$ matrix known as the factor loadings and D_i is a diagonal matrix $(i = 1, \ldots, g)$.

In practice, there is often the need to reduce further the number of parameters in the factor analysis model described above. Further restrictions can be applied to the specification of the elements of the component-covariance matrices. To this end, Baek and McLachlan [12] proposed the use of the so-called mixtures of common factor analyzers which uses common component-factor loadings that specifies the distribution of the feature vector Y_j conditional on its membership of component i as

$$Y_j = BU_{ij} + e_{ij}, \tag{9}$$

where B is a $p \times q$ matrix of common factor loadings and $U_{i1}, \ldots, U_{in}$ are distributed independently $N(0, I_q)$. It was so-named mixtures of common factor analyzers (MCFA) since the matrix B of factor loadings in (9) is common to all g components before the transformation of the factors U_{ij} to be white noise. This approach considerably reduces further the number of parameters in the MFA model. Moreover, it allows the data to be easily visualized in low dimensions by plotting the corresponding (estimated) factor scores.

Similar to the normal mixture model, the mixture of (normal) factor analyzers model is not robust to outliers since it uses the normal distribution for the errors and factors. In Baek and McLachlan [13], the so-called mixture of t-analyzers (MCtFA) was introduced in an attempt to make the model less sensitive to outliers, which replaces the normal distribution in the MCFA model with its more robust version— the multivariate t-distribution.

4   Dimension Reduction

When it is computationally feasible to fit the factor models as defined in the previous section, dimension reduction is effectively being done as part of the primary analysis. However, for many datasets in data mining applications, the number of variables p will be too large to fit these models directly without first performing some form of dimension reduction.

4.1   EMMIX-GENE

In the context of clustering, McLachlan et al. [14] proposed a three-step procedure (called EMMIX-GENE), in which in the first step the variables are considered individually by performing a test of a single t-component distribution versus a mixture of two t-components for each variable. Variables found not to be significant according to this test are discarded. Then on the second step, the retained variables (after appropriate normalization) are clustered into groups on the basis of Euclidean distance. Finally, on the third step, the observations can be clustered by the fitting of mixtures of normal or t-distributions or factor analyzers to representatives of the groups of variables.

Chan and Hall [15] have considered a method for variable selection where the variables are considered individually by performing a nonparametric mode test to assess the extent of multimodality.

4.2 Projection Methods

Another approach to dimension reduction is projection, which is very popular. For example, principal component analysis (PCA) is commonly used as a first step in the analysis of huge datasets, prior to a subsequent and more indepth rigorous modelling of the data. A related projection method is matrix factorization, which is described briefly in the next section.

4.3 Matrix Factorization

In recent times, much attention has been given to the matrix factorizations approach for dimension reduction. With this approach, the data matrix A as defined by (1) can be expressed as

$$A = A_1 A_2,$$ (10)

where the matrices A_1 and A_2 is a $n \times q$ matrix and a $q \times n$ matrix and, respectively. Here q is chosen to be much smaller than p. Given a specified value of q, the matrices A_1 and A_2 are chosen to minimize the function

$$\|A - A_1 A_2\|^2.$$ (11)

Dimension reduction is effected by replacing the data matrix A by the solution $\hat{A}_1$ for the factor matrix A_1, where the jth row of $\hat{A}_1$ gives the values of the q metavariables for the jth sample. This implies the original p variables are replaced by q metavariables. In the case where the elements of A are nonnegative, we can restrict the elements of A_1 and A_2 to be nonnegative. This approach is known as nonnegative matrix factorization (NMF) in the literature [16, 17]. We shall call the general approach where there are no constraints on A_1 and A_2, general matrix factorization (GMF).

The classic method for factoring the data matrix A is singular-value decomposition (SVD); see Golub and van Loan [18]. It follows from this theorem that the value of A_1 that minimizes (11) over the set of all $n \times q$ matrices of rank q is given by the matrix whose rows are the the eigenvectors corresponding to the q largest eigenvalues of $A^T A$.

Note that with SVD, effectively PCA, it imposes orthogonality constraints on the rows of the matrix A_1. Thus this ignores the possible nonindependence of the variables measured, for example, the genes in the biological processes under study; see, Kossenkov and Ochs [19]. On the other hand, the GMF approach has no constraints on C_2. Thus it provides a factorization of A into a low-dimensional subspace without any orthogonality constraints on its basis vectors. Hence, GMF is more flexible for modelling, for example, biological behaviour in which the gene signatures overlap. In contrast, due to its orthogonality constraints, PCA is overly restrictive for such data and is thus not suitable for the task of isolating gene signatures that have appreciable overlap. Furthermore, PCA is based on finding the directions of maximum variance, but the sample covariance matrix may provide misleading estimates where the number of variables p is much greater than the number n of observations [20].

Nikulin and McLachlan [21] have developed a very fast approach to the general matrix factorization (2), using a gradient-based algorithm that is applicable to an arbitrary (differentiable) loss function. Witten et al. [22] and Nikulin and McLachlan [23] have considered a penalized approach to PCA in order to provide sparse solutions.

5   Flow Cytometry Data

Data from a flow cytometry sample comprise parallel measurements of multiple physical, chemical and biological characteristics and properties of each individual cell. Briefly, cells in a sample are stained with a number of fluorophore-conjugated antibodies (or markers) before they are sequentially passed through laser beams in a flow cytometer that excite the fluorochromes. The light that emerges from each cell is captured and quantitated by a series of fluorescence detectors. These measurements are collected at a high speed, typically several thousands of cells per second, thus generating rich and complex high-resolution data in a high-throughput manner. In flow cytometric analysis, these simultaneous measurements of multiple fluorescent intensities can be used to study the differential expression of different surface and intracellular proteins, enabling biologists to study a variety of biological processes at the cellular level.

Traditionally, data are analyzed manually by trained analysts. The data are sequentially projected onto a series of bivariate plots and regions (gates) of interests are identified by the analyst manually, a process known as gating. With the advancement in technology allowing over 20 markers to be measured simultaneously, this has surpassed the practical feasibility of conventional manual analysis. Due to this and the rather the subjective nature of this approach, much recent efforts have focused on the development of computational tools to automate the gating process and to perform further analysis of big flow cytometry data including cluster alignment and supervised classification; see Aghaeepour et al. [24] for a recent account.

Mixture models provide a convenient and formal framework for modelling these complex and multimodal data. Under the mixture model framework, the flow cytometry data can be conceptualized as a mixture of populations, each of which consists of cells with similar expressions. The distribution of these cell populations can be characterized mathematically by a parametric density. Thus, the task of cell population identification translates directly to the classical problem of multivariate model-based clustering, where each cell population corresponds to a cluster of points with similar fluorescence intensities in the multidimensional space of markers.

5.1 Non-elliptical Clusters

It is well known that data generated from flow cytometric studies are often asymmetrically distributed, multimodal, as well as having longer and/or heavier tails than normal. To accommodate this, several methods adopted a mixture of mixtures approach where a final cluster may consist of more than one mixture component; see, for example, the procedures SWIFT [25], HDPGMM [26] and ASPIRE [27]. Another procedure, flowClust [28, 29], applied data transformation techniques to handle asymmetric clusters prior to using traditional mixture models. Further efforts have focused on the application of mixture models with skew distributions as components to enable a single component distribution to correspond to a cluster [4, 5, 30]. To this end, both FLAME [5] and JCM [4] adopt component densities that are a skew version of the t-distribution, which can be expressed as

$$f(y; \theta) = 2\, t_p(y; \mu, \Sigma, v)\, T_1\left(\frac{q}{\lambda}; 0, 1, v + p\right),\tag{12}$$

where $q = \delta^T \Sigma^{-1}(y - \mu)\sqrt{\frac{v+p}{v+d(y)}}$, $\lambda^2 = 1 - \delta^T \Sigma^{-1}\delta$, $d(y) = (y - \mu)^T\Sigma^{-1}(y - \mu)$ and δ is p-dimensional vector of skewness parameters. In (12), $t_p(y; \mu, \Sigma, v)$ denotes the p-variate t-density with location μ, scale matrix Σ and degrees of freedom v, and T_p denotes its corresponding distribution function. The t-distribution allows for heavier tails than the normal distribution, thus providing a more robust approach to traditional normal mixture modelling. The skew version (12) of the t-distribution, with an additional vector of skewness parameters δ, allows for non-symmetric distributional shapes. This version after an appropriate transformation is equivalent to the model proposed by Azzalini and Capitanio [31]. In recent years, many different versions of the skew t-distribution have been proposed; see [32] for an overview on this topic. Parameter estimation for various versions of skew normal and skew t-mixture models can be carried out using the EM algorithm [33, 34].

5.2 FLAME Procedure

The FLAME procedure was developed as a platform for automated gating (or clustering) and matching of flow cytometric data across samples and classes. With FLAME, each sample of cells can be effectively modelled by a mixture of skewed and heavy-tailed multivariate distributions. Segmentation of cells then corresponds to clustering based on the fitted skew mixture model. In the case of multiple samples (from the same cohort), the problem of matching cell populations across samples is handled by FLAME using a two-stage approach by undertaking Partitioning Around Medoids (PAM) and bipartite matching (BP). Firstly, a template consisting of the location of meta-clusters is constructed using PAM, based on the modes of all the populations identified by the mixture model in each sample. Subsequently, the population modes in each sample are matched to the template meta-clusters using a BP algorithm, which takes into account the proportions of cells in each population. This latter step allows extra or missing populations to be detected, and spurious populations to be recombined. The post-hoc steps are applicable not only to within-class matching of populations, but can also be directly applied to match meta-clusters across different classes.

5.3 JCM Procedure

In addition to the problem of clustering and matching of cell populations, the unsupervised classification (clustering) of unlabelled samples presents another challenging task in the pipeline of flow cytometric data analysis. While FLAME was primarily designed for the first two task, no overall parametric template was formed to describe the characteristics of the entire batch or class of samples and no algorithm was provided for the classification of unlabelled samples. Further, the matching of populations across samples is performed in a post-hoc fashion independent of the clustering step, and the accuracy of the PAM is sensitive to the amount of inter-sample variation in a batch.

Pyne et al. [4] proposed the multi-level JCM framework to address these concerns by operating on an entire batch of samples together. This enables the clustering and matching of populations to be performed simultaneously. Similar to FLAME, the JCM procedure model each populations with a component distribution. However, JCM introduced an additional higher level to link the individual mixture models from each sample to an overall parametric template. It is achieved by adopting a random-effects model (REM) to handle the inter-sample variations, where every sample in a given batch can be viewed as an instance of the batch template possibly transformed with a flexible amount of variation. An advantage of this approach is that each cluster in a sample is automatically registered with respect to the corresponding cluster of the template. A further advantage is the availability of an overall parametric template for the batch, which facilitates quantitative comparisons across different batches of samples. It also provides a basis for the allocation of an unlabelled sample to a class, where the batches represent training data from each different class. This latter task is handled by JCM by comparing the Kullback–Leibler (KL) divergence from the class template [35, 36]. Specifically, once a mixture model is fitted to an unlabelled sample, its label is predicted to be the class associated with the smallest KL divergence.

5.4 Clustering and Matching of Flow Cytometric Samples

To demonstrate the FLAME and JCM procedures in clustering and aligning cell populations in a batch of samples, we consider the diffuse large B-cell lymphoma (DLBCL) benchmark dataset from the flowCAP-I challenge (Aghaeepour et al., 2013). For this illustration, we use the subset of 16 samples which were manually analyzed and were determined to have the same number of clusters. These samples were derived from the lymph nodes of patients diagnosed with DLBCL, each stained with three markers—CD3, CD5 and CD19. Together with the forward scatter and side scatter measurements, each sample is represented by a $n \times p$ expression matrix with $p = 5$ and the number cells, n, ranging from 3000 to 25000.

Focusing first on the gating of these samples, we applied FLAME and JCM to cluster the 16 DLBCL samples. The FLAME procedure models each sample individually by fitting a mixture of skew t-distributions. With JCM, a template is created across the batch of 16 samples and each sample is linked to the template through an affine transformation. A skew t-mixture model is then fitted to each sample via the EM algorithm, using the JCM template as the initial values. The effectiveness of JCM and FLAME can be evaluated by comparing their clustering results with the results given by manual gating. The misclassification rate (MCR) and F-measure values for each sample are reported in Table 1 of Pyne et al. [4]. Note that a lower MCR and a higher F-measure indicate a closer match between the labels given the algorithm and those given by manual gating. The reported average MCR for JCM and FLAME across the entire batch of samples was 0.0711 and 0.1451, respectively. The corresponding average F-measures were 0.9403 and 0.9021. They compare favourably with the results for competing algorithms such as HDPGMM, flowClust and SWIFT, which obtained an average MCR of 0.2038, 0.1393 and 0.1454, respectively. These three algorithms were applied to this dataset as follows. The HDPGMM procedure fits a hierarchical mixture of Gaussian distributions to the dataset, where all samples share the same model parameters except for the mixing proportions. Hence, this model assumes that the location and shape of the cell populations are identical across all samples, and the weight or proportion of these populations is the only source of inter-sample variations. For flowClust and SWIFT, their models were applied to each sample individually.

In the case of alignment of cell populations across samples, FLAME can match the fitted component densities with the post-hoc steps described in Sect. 5.2. Starting from the models fitted to each samples (Fig. 1a) as produced by the clustering step above, FLAME runs PAM on the modes of these component densities to obtain

Table 1 Average MCR of various algorithms on the DLBCL dataset

Algorithm	JCM	FLAME	HDPGMM	flowClust	SWIFT
AMCR	0.0711	0.1451	0.2038	0.1151	0.3946

meta-clusters (Fig. 1b). The populations in each sample are then matched to these meta-clusters by undertaking BP (Fig. 1c). With JCM, the clustering and alignment of populations are performed simultaneously through the fitting of a JCM template (Fig. 1d).

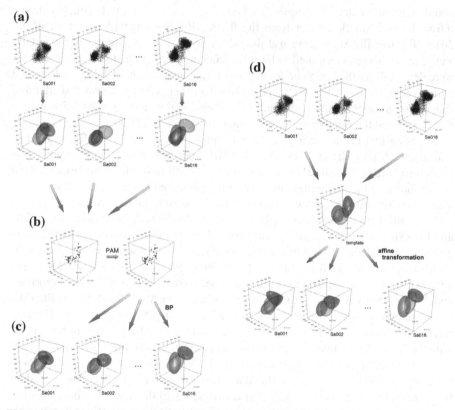

Fig. 1 Clustering and Matching of 16 DLBCL samples by FLAME and JCM. **a** FLAME clustering: a skew t-mixture model is fitted to each sample individually. **b** FLAME meta-clustering: the modes of all clusters from each sample were pooled and clustered using PAM to construct a template of meta-clusters. **c** FLAME populations matching: clusters from each sample were matched to meta-clusters using BP. **d** JCM: clustering and matching are performed simultaneously by fitting a random-effects model where each sample is linked to a template model through an affine transformation

5.5 Classification of Flow Cytometric Samples

In a recent study on germinal centre (GC) lymphoma, Craig et al. [37] studied cellular hierarchies that may help distinguish germinal centre hyperplasia (GC-H) from germinal centre B-cell lymphoma (GC-L). Lymphoid tissue biopsy specimens from 100 individuals were analyzed using flow cytometry with markers CD45, CD20, kappa, lambda, CD19, CD5, CD10 and CD38. Several other measurements were also recorded for each cell, including forward and side scatter measurements, resulting in a collection of 100 data files, each with 15 variables and around 30,000 observations. These samples were also manually analyzed to obtain training data. There were 48 cases of GC-L and 52 cases of GC-H. For our illustration, we allow JCM to train on 70 samples (consisting 37 cases of GC-L and 33 cases of GC-H), and evaluate the classification performance of JCM on the remaining 30 samples.

We applied the JCM procedure and six other algorithms (ASPIRE, Citrus, flowMatch, flowPeaks, SWIFT and HDPGMM) to this dataset. Since the latter three algorithms do not provide a strategy for sample classification, we follow a similar approach to Dundar et al. [27] where a support vector machine (SVM) classifier is built based on the mixing proportions of the template model applied to each sample. For SWIFT and flowPeaks, the template model is computed by pooling all samples together. For HDPGMM, a consensus model is obtained by modal clustering. With JCM, ASPIRE, flowPeaks, flowMatch, SWIFT and HDPGMM, their templates identified 9, 2, 11, 4, 8 and 16 clusters, respectively. Table 2 reports some common measures of classification performances for these algorithms on the GC dataset, including sensitivity, specificity, accuracy, precision and F-measure. Among these algorithms, JCM achieved the highest F-measure value for this dataset, as well the highest accuracy. Both JCM and HDPGMM yield perfect specificity and precision. On comparing the scores or measures used by the classifiers of these various methods (Fig. 2), the classification score adopted by JCM has the best classification performance for this dataset. The classification scores for each of the methods is calculated as follows. In Fig. 2a, the KL score of JCM is calculated as the KL distance from the GC-L class over the sum of the KL distance from the GC-L and GC-H classes. Thus, a higher score means that the model fitted to the sample is closer to the GC-L template. For SWIFT, flowPeaks and HDPGMM (Fig. 2b, d, f respectively), the decision values given by the SVM classifier are shown. In Fig. 2c, a classification score for the Citrus procedure can be obtained by calculating the proportion of Citrus classifiers that predicted GC-L for a sample. Figure 2e shows the scores computed by flowMatch; see Azad et al. [38] for details.

6 Summary and Conclusions

We have considered how mixture models can be applied to large datasets involving many observations and/or variables. Two real examples in flow cytometry are pre-

Table 2 Classification results of various algorithms on the GC dataset

	Sensitivity	Specificity	Accuracy	Precision	F-measure
JCM	0.74	1.00	0.83	1.00	0.85
SWIFT	0.89	0.64	0.80	0.81	0.74
ASPIRE	0.74	0.55	0.67	0.74	0.63
Citrus	0.58	0.64	0.60	0.73	0.61
flowPeaks	0.52	0.64	0.57	0.71	0.58
flowMatch	0.68	0.45	0.60	0.68	0.55
HDPGMM	0.32	1.00	0.57	1.00	0.48

sented. For high-dimensional datasets where traditional normal and t-mixture models cannot be fitted directly, mixtures of factor models provide a computationally less demanding alternative, whereby the full observation vector is modelled by projecting onto low-dimensional spaces. Further reduction in the number of parameters to be estimated can be achieved using factor models with common factor loadings. In cases where even factor models may not be computationally feasible, dimension reduction techniques such as variable selection, projections and matrix factorization, can be applied prior to the fitting of mixtures of factor analyzers.

Recent advances in flow cytometry technology have enabled rapid high-throughput measurements to be taken, generating big and high-dimensional datasets for analysis. This has sparked the development of various computational tools for automated analysis of flow cytometry samples. Among these methods, mixture models have been widely adopted as the underlying mechanism for characterizing the heterogeneous cell populations in these cytometry samples; see, for example the procedures FLAME [5], SWIFT [25], flowClust [28], flowPeaks [39], HDPGMM [26], ASPIRE [27], JCM [4] and the articles [30, 40–42]. Observing that these clusters of cells in flow cytometry samples are typically not elliptically symmetric and may contain outliers, one approach is to consider the fitting normal mixture models with a large number of components and then subsequently merging some of the components. However, this entails the problem of trying to identify the correspondence between components and clusters. To tackle this, FLAME adopted mixture models with skew normal and skew t-component densities that can flexibly accommodate asymmetrically distributed clusters with tails that may be longer than normal. In the analysis of a cohort of flow cytometry data, consisting of multiple samples from different classes that may corresponds to different clinical outcomes, one is also interested in the classification of unlabelled samples to a predefined class. To this end, some of the above mentioned procedures have been extended to perform alignment of cell populations across multiple sample and the construction of a template to represent the characteristics of a batch of samples; for example, ASPIRE and JCM. This facilitates (supervised) classification of new samples. These two methods introduced random-effects to cater for inter-sample variations, with the class template of the former procedure based on hierarchical mixtures of normal distributions, while the lat-

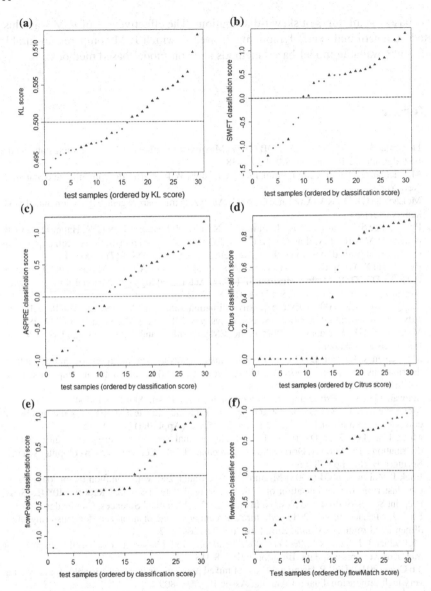

Fig. 2 Plots displaying the scores or measures used by different classifiers for the GC dataset. *Blue triangles* denote a GC-H sample, whereas *red dots* denote a GC-L sample. A *dash line* indicate the classification boundary used by a classifier. **a** (JCM) scatter plot of the KL score by JCM, given by the relative KL distance from the GC-L class. Samples corresponding to points that lie above the boundary (*dashed line*) are likely to be GC-L. **b** (SWIFT) scatter plot of the scores given by the SVM classifier. The threshold is optimized based on the training data. Note that samples are ordered in ascending order of SVM scores. **c** (ASPIRE) scatter plot of the SVM classifier. **d** (Citrus) scatter plot of the classifier score for each sample (based on the results of the 100 regression classifiers selected by Citrus). Samples corresponding points that lie above the boundary are classified as GC-H. **e** (flowPeaks) scatter plot of the scores given by the SVM classifier. **f** (flowMatch) scatter plot of scores

ter is based on mixtures of skew t-distributions. The effectiveness of JCM was illustrated on a germinal centre lymphoma dataset, in which JCM compares favourably with other available model-based methods and non-model-based methods.

References

1. Dempster AP, Laird NM, Rubin DB (1977) Maximum likelihood from incomplete data via the EM algorithm. J R Stat Soc, Ser B 39:1–38
2. McLachlan GJ, Peel D (2000) Finite mixture models. Wiley Series in Probability and Statistics, New York
3. McLachlan GJ, Do KA, Ambroise C (2004) Analyzing microarray gene expression data. Hoboken, New Jersey
4. Pyne S, Lee SX, Wang K, Irish J, Tamayo P, Nazaire MD, Duong T, Ng SK, Hafler D, Levy R, Nolan GP, Mesirov J, McLachlan GJ (2014) Joint modeling and registration of cell populations in cohorts of high-dimensional flow cytometric data. PLOS ONE 9(7):e100334
5. Pyne S, Hu X, Wang K, Rossin E, Lin TI, Maier LM, Baecher-Allan C, McLachlan GJ, Tamayo P, Hafler DA, De Jager PL, Mesirow JP (2009) Automated high-dimensional flow cytometric data analysis. Proc Natl Acad Sci USA 106:8519–8524
6. Li JQ, Barron AR (2000) Mixture density estimation. In: Solla SA, Leen TK, Mueller KR (eds) Advances in neural information processing systems. MIT Press, Cambridge, pp 279–285
7. McLachlan GJ, Krishnan T (2008) The EM algorithm and extensions, 2nd edn. Wiley-Interscience, Hokoben
8. McLachlan GJ, Peel D (1998) Robust cluster analysis via mixtures of multivariate t-distributions. In: Amin A, Dori D, Pudil P, Freeman H (eds) Lecture notes in computer science, vol 1451. Springer, Berlin, pp 658–666
9. Schwarz G (1978) Estimating the dimension of a model. Ann Stat 6:461–464
10. McLachlan GJ (1987) On bootstrapping the likelihood ratio test statistic for the number of components in a normal mixture. J R Stat Soc Ser C (Appl Stat) 36:318–324
11. McLachlan GJ, Peel D (1998) Robust cluster analysis via mixtures of multivariate t-distributions. In: Amin A, Dori D, Pudil P, Freeman H (eds) Lecture notes in computer science. Springer, Berlin, pp 658–666
12. Baek J, McLachlan GJ (2008) Mixtures of factor analyzers with common factor loadings for the clustering and visualisation of high-dimensional data. Technical Report NI08018-SCH, Preprint Series of the Isaac Newton Institute for Mathematical Sciences, Cambridge
13. Baek J, McLachlan GJ (2011) Mixtures of common t-factor analyzers for clustering high-dimensional microarray data. Bioinformatics 27:1269–1276
14. McLachlan GJ, Bean RW, Peel D (2002) A mixture model-based approach to the clustering of microarray expression data. Bioinformatics 18:413–422
15. Yb Chan (2010) Hall P. Using evidence of mixed populations to select variables for clustering very high dimensional data. J Am Stat Assoc 105:798–809
16. Lee DD, Seung HS (1999) Learning the parts of objects by non-negative matrix factorization. Nature 401:788–791
17. Donoho D, Stodden V (2004) When does non-negative matrix factorization give correct decomposition into parts? In: Advances in neural information processing systems, vol 16. MIT Press, Cambridge, MA, pp 1141–1148
18. Golub GH, van Loan CF (1983) Matrix computation. The John Hopkins University Press, Baltimore
19. Kossenkov AV, Ochs MF (2009) Matrix factorization for recovery of biological processes from microarray data. Methods Enzymol 267:59–77
20. Johnstone IM, Lu AY (2009) On consistency and sparsity for principal components analysis in high dimensions. J Am Stat Assoc 104:682–693

21. Nikulin V, McLachlan G (2009) On a general method for matrix factorisation applied to supervised classification. In: Chen J, Chen X, Ely J, Hakkani-Tr D, He J, Hsu HH, Liao L, Liu C, Pop M, Ranganathan S (eds) Proceedings of 2009 IEEE international conference on bioinformatics and biomedicine workshop. IEEE Computer Society, Washington, D.C. Los Alamitos, CA, pp 43–48

22. Witten DM, Tibshirani R, Hastie T (2009) A penalized matrix decomposition, with applications to sparse principal components and canonical correlation analysis. Biostatistics 10:515–534

23. Nikulin V, McLachlan GJ (2010) Penalized principal component analysis of microarray data. In: Masulli F, Peterson L, Tagliaferri R (eds) Lecture notes in bioinformatics, vol 6160. Springer, Berlin, pp 82–96

24. Aghaeepour N, Finak G (2013) The FLOWCAP Consortium, The DREAM Consortium. In: Hoos H, Mosmann T, Gottardo R, Brinkman RR, Scheuermann RH (eds) Critical assessment of automated flow cytometry analysis techniques. Nature Methods 10:228–238

25. Naim I, Datta S, Sharma G, Cavenaugh JS, Mosmann TR (2010) Swift: scalable weighted iterative sampling for flow cytometry clustering. In: IEEE International conference on acoustics speech and signal processing (ICASSP), 2010, pp 509–512

26. Cron A, Gouttefangeas C, Frelinger J, Lin L, Singh SK, Britten CM, Welters MJ, van der Burg SH, West M, Chan C (2013) Hierarchical modeling for rare event detection and cell subset alignment across flow cytometry samples. PLoS Comput Biol 9(7):e1003130

27. Dundar M, Akova F, Yerebakan HZ, Rajwa B (2014) A non-parametric bayesian model for joint cell clustering and cluster matching: identification of anomalous sample phenotypes with random effects. BMC Bioinform 15(314):1–15

28. Lo K, Brinkman RR, Gottardo R (2008) Automated gating of flow cytometry data via robust model-based clustering. Cytometry Part A 73:312–332

29. Lo K, Hahne F, Brinkman RR, Gottardo R (2009) flowclust: a bioconductor package for automated gating of flow cytometry data. BMC Bioinform 10(145):1–8

30. Frühwirth-Schnatter S, Pyne S (2010) Bayesian inference for finite mixtures of univariate and multivariate skew-normal and skew-t distributions. Biostatistics 11:317–336

31. Azzalini A, Capitanio A (2003) Distribution generated by perturbation of symmetry with emphasis on a multivariate skew t distribution. J R Stat Soc Ser B 65(2):367–389

32. Lee SX, McLachlan GJ (2013) On mixtures of skew-normal and skew t-distributions. Adv Data Anal Classif 7:241–266

33. Lee S, McLachlan GJ (2014) Finite mixtures of multivariate skew t-distributions: some recent and new results. Stat Comput 24:181–202

34. Lee SX, McLachlan GJ (2016) Finite mixtures of canonical fundamental skew t-distributions: the unification of the unrestricted and restricted skew t-mixture models. Stat Comput. doi:10.1007/s11222-015-9545-x

35. Lee SX, McLachlan GJ, Pyne S (2014) Supervised classification of flow cytometric samples via the joint clustering and matching procedure. arXiv:1411.2820 [q-bio.QM]

36. Lee SX, McLachlan GJ, Pyne S. Modelling of inter-sample variation in flow cytometric data with the joint clustering and matching (JCM) procedure. Cytometry: Part A 2016. doi:10.1002/cyto.a.22789

37. Criag FE, Brinkman RR, Eyck ST, Aghaeepour N (2014) Computational analysis optimizes the flow cytometric evaluation for lymphoma. Cytometry B 86:18–24

38. Azad A, Rajwa B, Pothen A (2014) Immunophenotypes of acute myeloid leukemia from flow cytometry data using templates. arXiv:1403.6358 [q-bio.QM]

39. Ge Y, Sealfon SC (2012) flowpeaks: a fast unsupervised clustering for flow cytometry data via k-means and density peak finding. Bioinformatics 28:2052–2058

40. Rossin E, Lin TI, Ho HJ, Mentzer S, Pyne S (2011) A framework for analytical characterization of monoclonal antibodies based on reactivity profiles in different tissues. Bioinformatics 27:2746–2753

41. Ho HJ, Lin TI, Chang HH, Haase HB, Huang S, Pyne S (2012) Parametric modeling of cellular state transitions as measured with flow cytometry different tissues. BMC Bioinform. 2012. 13:(Suppl 5):S5

42. Ho HJ, Pyne S, Lin TI (2012) Maximum likelihood inference for mixtures of skew student-t-normal distributions through practical EM-type algorithms. Stat Comput 22:287–299

An Efficient Partition-Repetition Approach in Clustering of Big Data

Bikram Karmakar and Indranil Mukhopadhayay

Abstract Addressing the problem of clustering, i.e. splitting the data into homogeneous groups in an unsupervised way, is one of the major challenges in big data analytics. Volume, variety and velocity associated with such big data make this problem even more complex. Standard clustering techniques might fail due to this inherent complexity of the data cloud. Some adaptations are required or demand for novel methods are to be fulfilled towards achieving a reasonable solution to this problem without compromising the performance, at least beyond a certain limit. In this article we discuss the salient features, major challenges and prospective solution paths to this problem of clustering big data. Discussion on current state of the art reveals the existing problems and some solutions to this issue. The current paradigm and research work specific to the complexities in this area is outlined with special emphasis on the characteristic of big data in this context. We develop an adaptation of a standard method that is more suitable to big data clustering when the data cloud is relatively regular with respect to inherent features. We also discuss a novel method for some special types of data where it is a more plausible and realistic phenomenon to leave some data points as noise or scattered in the domain of whole data cloud while a major portion form different clusters. Our demonstration through simulations reveals the strength and feasibility of applying the proposed algorithm for practical purpose with a very low computation time.

1 Introduction

The fun with the concept of big data is that while its existence is undoubted, there cannot be any formal way to define it in the language of mathematics. It has been argued that the major characteristics of big data are volume, variety, velocity and

B. Karmakar
Department of Statistics, The Wharton School, University of Pennsylvania,
Jon M Houston Hall, Philadelphia, PA 19104, USA
e-mail: bikramk@wharton.upenn.edu

I. Mukhopadhayay (✉)
Human Genetics Unit, Indian Statistical Institute, Kolkata 700 108, India
e-mail: indranil@isical.ac.in

© Springer India 2016
S. Pyne et al. (eds.), *Big Data Analytics*, DOI 10.1007/978-81-322-3628-3_5

complexity [10]. These make big data analysis challenging and also distinguish it from classical data analysis. Thinking back, big data is not the only member of the class of concepts in data science that is not defined in mathematically precise manner. Clustering is another such technique which does not have any axiomatic definition. The task of clustering a data would be familiar to everyone but making the problem precise seems impossible [13].

Broadly, clustering has been viewed as a tool to achieve one of following three goals: (1) data exploration, (2) confirmation of hypothesis, and (3) preprocessing data to felicitate some objectives in mind. We discuss below these three aspects of clustering in the context of big data. This discussion should allow the reader to comprehend the practical use of clustering ideology in big data analytics.

Data exploration—Finding hidden patterns in the data is the first and foremost objective of data analysis. If we want to put this idea in perspective of big data one would be very ambitious about extracting more and more interesting patterns. This optimism is a powerful force in driving interesting researches on big data analytics. If one thinks about the problem a little more carefully, even putting aside the trouble of actually being able to handle the data, it would become evident that just an abundance of data is not necessarily the best thing. What is more important is the quality of the data and our ability of assess that quality through stable methods to extract quantitative information out of such data. The idea of big data itself compromises with the quality of the data. Big data accumulated from heterogeneous sources, sometimes in an uncontrolled manner, introduces enormous amount of noise. This noise may be due to measurement error, presence of outliers, significant amount of missing observations, etc. Even in more hopeful scenarios like genetic data or fMRI data where we have a control on data generation process, noise in the data is introduced as a curse of dimensionality or due to other reasons. In such a case we would be looking for some small set of genes or pixels, respectively, while all other genes or pixels are essentially noise.

Conformation of hypothesis—This is the exactly as it says: validating a precomposed hypothesis based on the data. Specific to clustering context, if we expect that there is a particular clustering in the data, then we would want to check whether such an expectation is consistent with observed data. This is actually a pretty generous expectation from big data analysis. We should be able to provide a satisfactory answer to this problem. Once validated, such hypotheses could lead to follow up analysis of the data or provide a guidance on further experimental data collection.

To make the problem more precise, suppose that for a dataset collected on individual level, it is hypothesized that a set of features determines a cluster according to a relation $\mathbf{x} \in S$. Hence, our hypothesis says, individuals with $\mathbf{x} \in S$ have some sort of homogeneity among themselves. A straightforward way to validate such hypothesis is to consider a sample of paired individuals, $\mathscr{G}_S$, each pair having $\mathbf{x} \in S$ and compare them with a bunch of pairs of individuals, $\mathscr{G}$ where the individuals do not have any restriction on feature $\mathbf{x}$. In data science there are plenty of summary statistics one can think of to carry out such comparisons. Let us consider a simple one. Let V_S and V be the total variability of feature differences for features other than $\mathbf{x}$ for the groups $\mathscr{G}_S$ and $\mathscr{G}$, respectively. Deicide that the data do confirm the hypothesis if $V \gg V_S$. The blessing of big data is that if we repeat this decision making on quite a few number of samples, then with a majority vote of these repeated decisions we shall have a very satisfactory decision rule. The challenge then remains to implement an efficient algorithm to carry out this decision making.

On the first look the method discussed above may seem incoherent to usual clustering methods. This is because while in clustering based on different similarity or distance measures we aim to extract homogeneous groups, here the aim is slightly different. Given a feature that is expected to create a homogeneous group we are asking, if such an expectation is consistent with our similarity or distance measure. This practical point of view is not always addressed in clustering literature, especially in the context of big data analytics.

Clustering as preprocessor—Even if the data are intended to be used for some different purposes other than exploration or confirmation, clustering can still be used as a preprocessor so that it would make later steps more feasible. Let us put this idea in the context of the trend of big data analytics. Computation on big data is carried out on clusters of machines where in each machine a 'part' of the data is analysed. These parts are determined based on either randomization or based on sources or using a mapping that distributes the job into multiple machines. Thus the issue of computational ability should be kept in mind while distributing the data to different machines. Now suppose our objective is variable selection. Then to avoid the challenge of spurious correlation of the data, clusters of time gapped data points is a much better choice than random clusters. Further, such preprocessing leads us to understand the nature of data in different sized bins so as to apply appropriate statistical techniques and analytical tools. This would also reveal any significant correlation among the variables.

In this article we are going to discuss in details about clustering as a tool of data exploration in big data analytics. The arrangement of the sections is as follows. In Sect. 2, we shall state a formalism of clustering big data. Section 3 is dedicated to elaborate on the primary classes of big data and the distinction of clustering principles among them. A comprehensive literature review of big data clustering algorithms is presented in Sect. 4. In Sect. 5 two meta algorithms are presented for big data clustering. A broad range of simulation studies are carried out in Sect. 6 for the proposed meta algorithms.

2 Formulation of Big Data Clustering

Conceptually, clustering of big data or a very large dataset although same to that of clustering of a moderately sized dataset in principle, they vary in many aspects while constructing the clusters. One of the basic problems is due to the sheer volume or size of big data. Usually it exceeds the RAM of a computer and hence is extremely difficult to apply in one go. So summarizing the results from a clustering algorithm requires utmost attention; otherwise huge amount of information might be lost and hence the downstream analysis would be meaningless or less sensitive and less fruitful.

To tackle the volume of big data, partitioning is one option but the method of such partitioning is important as it can be done in various ways. The easiest in theory is to divide the dataset at random but in many situations data may come from multiple sources thus adding complexity to the data cloud. Applying clustering algorithm to a single partition is always manageable but summarizing the results requires precise method and plan so that memory space could be used in an optimum way at the same time preserving the information obtained thus far. Use of identifier or data labels to the actual dataset instead of storing the whole data might be a good option towards fulfilling this purpose.

Another problem is that there might be a chance that one true cluster is divided into two clusters due to partitioning thus increasing the number of clusters. This might mislead the downstream analysis of the data. So there should be a check to see the nature of clusters that are obtained through different machines in parallel. After checking, a global clustering labelling should be allotted to each data point. Thus the summarization process needs to be revisited based on the output of the clustering algorithm adapted to the specificity of big data.

Result of clustering would be informative in finding the hidden pattern that includes information about the correlation structure among the data points. Moreover, this provides an overview of the space covered by points in a cluster and the map of spaces for different clusters. This would naturally generate some idea about the future analysis that can be done using the same dataset. Thus proper preprocessing during clustering and retaining the information is essential to get more insight to the big data set.

3 Principles of Big Data Clustering

Most implementation of clustering algorithms that are considered for big data analytics are, in a sense, applicable to large data. As mentioned above the obvious difference between big data and simply a large or high dimensional data set is that big data adds a level of complexity of data storage and increase in volume of data with time. This complexity may be due to accumulation of data from multiple sources, among other reasons. Consequently a clustering algorithm should operate only locally and update the results sequentially.

In a big data, the volume of the data is one of its primary characteristics and concerns. When we are faced with the problem of variety in the data due to heterogeneous sources, we are left hopeless in any sort of use of clustering algorithms. Let us consider the size aspect of a big data for the moment. Even if the volume is somewhat inconceivable in practice, there are recognizable patterns in which the volume is accumulated. In a *structured* data, we can assert that volume is accumulated in one of the following two patterns.

Fixed Dimension—A large proportion of data generated in *automated* manner is essentially of fixed and very low dimension. These are mostly user data on various services. The service provider accumulates data on only a few variables of interest and the data accumulates its volume as users are served at a very high frequency. A specific example of such data is traffic data which collect information on the amount of time taken by a traveller (with a GPS device) to travel between geographical locations and a few other relevant covariates (e.g. mode of transportation, tweets with a certain hashtag, etc.).

The particular challenges in clustering such data can be characterized by (i) immobility and (ii) noise accumulation. (i) Immobility of the data means that once a dataset is collected it is unrealistic to expect that data would be transferred completely in another physical memory. Immobility is garnered since the ultra high frequency of data generation requires sequential storages. Practically it is unreasonable to expect to get all the data in one storage and process it. This imposes the restriction on a clustering algorithm for such a data be localized. This means that the algorithm processes the data on its own physical location without having it copied somewhere else entirely. (ii) Noise accumulation on the other hand is an even more serious threat to analyzing this kind of data. Noise in the data is undesirable especially because we have only a fixed dimension. If over time amount of noise gathered is more than informative data, then any clustering algorithm that fails to accommodate such scenario cannot hope to learn anything useful and would be no better (or even less) than a random partitioning.

High Dimension—For this class of big data the volume of the data increases more in terms of dimension than in its size. Broadly there can be two subclasses of such data. In one subclass, generally there is a cost of gathering data for each individual. Popular examples are gene expression data and medical imaging data where data of very high dimension are collected only on a few individuals. In another subclass of data, individual data are generated over time. One such example is consumer purchase data collected over time by some big vendor or retailer. While in the second subclass the number of individuals can grow as well; such growth (in number of customers) is expected to be much smaller than growth in dimension (information on purchases).

Interestingly, for this class of data we might be looking for either cluster of individuals or cluster of variables based on the subclass we are in. In the first of the two subclasses, we are aiming for groups of variables that have special patterns. The principle challenge is to find the most *relevant* groups/clusters. Thus all the variables (e.g. genes in a gene expression data) may not be of any relevance and may be termed as *"noise"* in the dataset. Immobility of the data is less serious problem in processing this type of data. A subsampling-based algorithm is suitable for analyzing such data. In the second subclass of data where we are gathering individual level data over time, we would be interested in grouping individuals based on their pattern. In such a scenario, clustering is more important as a preprocessing tool. Then challenge is posed by spurious correlation among the variables. Dimension reduction methods might work as an essential tool for clustering such data. Some of basic principles of clustering big data can be summarized in Table 1 below.

Table 1 A summary: principles for clustering big data based on different data classes

Structured data (no variety)

	Fixed dimension	High dimension	
		Huge dimension	Increasing number of Dimension
Source of volume	High frequency of observations	Dimension	Dimension
Example	traffic data, customer survey data	Gene expression data, medical imaging data	Consumer data
Objective/goal	Cluster individuals	Cluster relevant variables	Clustering individuals of similar pattern
Challenge	Immobility of data and noise accumulation	Too many irrelevant variables	Immobility of data and spurious correlation
Feature	Growing number of clusters	Irrelevant variables work as noise	
Principle	Localized algorithms, accommodate for noise, sequential updating	Subsampling-based algorithm	Localized algorithm, dimension reduction

Thus volume aspect provides us with quite a bit of insight into the challenges that a clustering algorithm must face in different situations. More important thing to note from the above discussion is that, a single clustering algorithm is not desired to work in all scenarios of big data.

Big data clustering principles thus broadly are one of the following three types: (1) Localized algorithm, (2) Subsampling-based algorithm and (3) Dimension Reduction (and Localized algorithm). In Sect. 4, we will discuss a few subsampling-based and dimension reduction-based clustering algorithms that exist in the literature. Unfortunately the principle of localized algorithm has failed to receive the desired attention. At this point we prefer first to elaborate more on this principle.

Localized Clustering Principle—As transpired from our discussion, localized algorithms are absolutely required when our big data problem in turn imposes data immobility issue. Any such algorithm would require a two step procedure. In the first step, we would extract clusters from a local data storage. Velocity in the data, say, data collected over 'days', then requires that this first step procedure be carried out in sequential manner. The second step which is more algorithmically challenging is to identify which clusters in the current 'day' are new ones and which ones are same as some of the clusters from previous 'days'. This step requires checking 'equality' of two clusters. Then we need to answer the following two questions: (1) what is the right measure to check for this 'equality', and (2) how do we implement this checking computationally efficiently and without moving too much of data around? In Sect. 5 we shall present a two-step version of good old K-means algorithm. This two-step procedure impressively cuts down the cost of computation and memory requirement.

Recall that in high dimensional data clustering of variables we are required to produce relevant clusters, e.g. most significant genes for a microarray gene expression data. This added requirement makes even basic clustering algorithms quite computationally expensive. Thus, in such a scenario if we adopt the two-step principle discussed above over a sampling-based algorithm, we might be quite efficient in achieving our goal. In Sect. 5, we will present an algorithm that is based on our ongoing work following this principle in a rigorous manner.

In the next section, we shall discuss the clustering algorithms available in the literature which have the hope to be useful in big data analytics.

4 Literature Review

The literature of clustering is flooded with many innovative methods that are applicable to data under several situations. Among them K-means and hierarchical clustering are very old and stably used in clustering of different types of data. We must note that here we only consider clustering of continuous data vectors and not other types of data that include categorical data. Computational efficiency is one of the most important features and to this end we find other clustering algorithms that mainly emphasize this aspect of clustering. However, at the end, cluster quality is the main issue that we should concentrate as the next level analysis solely depends on this quality. But the main question is whether these clustering algorithms can be applied with high degree of efficiency to the big data as well. Brief reviews of big data clustering by [3, 16] are worth mentioning in this context. They have correctly identified the difficulties and challenges for applying standard clustering algorithms to big data. Due to its sheer volume only, big data can make our lives difficult because the computational time is too high and we might not get reasonable clusters at all using a single computer. Moreover, even with a high end computer, the computation may be slowed down severely due to the tremendous volume of data that is usually kept in the RAM of the machine.

Thus high complexity and computational cost sometimes make the clustering of big data prohibitive. Even for K-means clustering, this creates problems when the number of data vectors is not so large compared to the dimension of each data point. Here big data puts more importance to the dimension rather than the size of data set. Hence in such situations scalability is the main challenge. Moreover some adaptation of these standard algorithms sometimes might cause substantial sacrifice in terms of cluster quality. Although multiple machine-based techniques can be proposed, these should not be compromised with the quality of clustering and should satisfy some optimum properties.

Unlike partitioning around memoids (PAM) [11], clustering large applications (CLARA) [11] decreases overall time complexity from $O(n^2)$ to a linear order. Moreover clustering large applications based on random sampling (CLARANS) [15] method is proposed to improve the efficiency over CLARA. Another method BIRCH [20] takes care of clustering of large data sets that are even larger than the memory size of the machine. Some density-based methods are available in the literature that search for connected dense regions in the sample space and identify areas that are heavily populated with data. Such methods viz, DBCURE [12], DBSCAN [4], Optics [2, 14], DENCLUE [6], etc. are mainly used to construct clusters with irregular shapes but are not suitable for high dimensional data set.

Hierarchical clustering mainly is of two types: divisive and agglomerative. Hierarchical divisive methods start with one single cluster and try to divide it into other clusters until each point is put in a different cluster. Hierarchical agglomerative methods work exactly in the opposite manner but the philosophy of considering each point as one cluster remains the same. Here it starts with each point and try to merge other points into a bigger cluster. Both these approaches use different types of distance measure between groups. Naturally hierarchical clustering have complexity as high as of order n^2 and hence not scalable, making it not at all appealing for big data clustering.

However, in big data clustering perspective, some proposals have been put forward using the computing power of a single machine and using multiple such machines. Parallel clustering and MapReduce-based clustering are the most popular ones. DBDC [1, 8] and ParMETIS [9] are some examples of multilevel partitioning algorithms that improve DBSCAN and METTIS, respectively, for parallel computation. MapReduce is a framework that is initially represented by Google and its open source version is known as Hadoop. MapReduce-based K-means clustering, known as PKmeans [21], is an adaptation of standard K-means algorithm that distributes the computation among multiple machines using MapReduce. The main objective of PKmeans is to speed-up and scale-up the computing process. Another version of K-means is ELM K-means [19] that is seen to be suited nicely for finding good quality clusters on less computation time thus showing a promise to its adaptability to big data clustering. Similarly MR-DBSCAN [5] is a scalable MapReduce-based version of DBSCAN that can be used in multiple machines.

Although clustering algorithm when applied in parallel, increases the scalability and speed, evaluation of such an application is needed for understanding its usefulness as compared to the application of a single clustering to the whole data, if possible. Thus results that come from different machines should be evaluated to see how the actual clusters look like; whether they really form "good" clusters in terms of distance-based measure between clusters and within clusters. With technological advances speeding up of computation would be more feasible and possible in future to a great extent, but statistical validation of the quality of clusters thus produced should be kept in mind and needs to be studied elaborately. Nowadays GPUs are much more powerful than CPUs and so focus is on developing appropriate algorithms and apply them in a distributed manner to increase the scalability and speed.

5 Clustering Meta Algorithm/Partition-Repetition Approach

We dedicate this section to present two algorithms for big data clustering. These two algorithms are motivated by our discussion in Sect. 3. However, our approaches can be extended easily to incorporate any other appropriate clustering engines. These proposal for generic algorithms are based on the need of analysis and nature of big data. It can be adapted for specific purpose with appropriate adjustments. Our proposed method is based on the philosophy which we call Partition-Repetition algorithm, as explained below.

Partitioning—Suppose that a dataset consists of n data points (e.g. genes). We randomly partition n points into $L(n)$ parts. If we can run a clustering algorithm on a data of size M, we partition the data into $L(n) \sim \lceil \frac{n}{M} \rceil$ parts. To reduce the error due to partitioning one may wish to repeat this procedure $R(n)$ times so that we are left with $M_0 = R(n) \times L(n)$ manageable datasets and apply the clustering algorithm of our choice on each such dataset. We store all clusters from M_0 datasets in an empty bucket $\mathscr{B}$. The elements in this bucket can be enumerated as

$$\mathscr{B} = \{C_1, C_2, \dots, C_{M_0}\}.$$

Combining—Consider the first member of the bucket C_1. We want to combine all members of the bucket with the similar pattern with that of C_1 and redefine it as a new collection. Let us consider a decision function $\pi(\cdot, \cdot)$ that maps $2^X \times 2^X$ to $\{0, 1\}$, where 2^X denotes the power set of X. If $\pi(C, \hat{C}) = 1$, then we decide whether the subsets C and $\hat{C}$ are of similar profile or they are of different profiles. With the help of this decision function, bucket elements are redefined as,

$$C_1 := C_1 \bigcup \left(\bigcup_{j=2}^{M_0} \mathscr{I}\left(C_j, \pi\left(C_1, C_j\right)\right) \right), \tag{1}$$

and

$$C_j := \mathscr{I}\left(C_j, 1 - \pi\left(C_1, C_j\right)\right) \quad for\ j = 2, 3, \dots, M_0. \tag{2}$$

Here, $\mathscr{I}(\cdot, \cdot)$ is a set valued function defined on $2^X \times \{0, 1\}$ as follows: $\mathscr{I}(C, 1) = C$ and $\mathscr{I}(C, 0) = \phi$. At the end of this redefinition step, we are again left with M_0 members in the bucket but now possibly some of them are empty. This process of redefining is carried out sequentially for $i = 2, 3, \dots, M_0 - 1$ as follows,

$$C_i := C_i \bigcup \left(\bigcup_{j=i+1}^{M_0} \mathscr{I}\left(C_j, \pi(C_i, C_j)\right) \right), \tag{3}$$

and

$$C_j := \mathscr{I}\left(C_j, 1 - \pi\left(C_i, C_j\right)\right) \quad for \; j = i + 1, \dots, M_0. \tag{4}$$

A necessary constraint on the decision function π is that $\pi(\phi, C) = \pi(C, \phi) = 0$ for any $C \subseteq X$.

We can denote this method as an operation on $\mathscr{B}$, say $\mathscr{B} := \mathbb{C}(\mathscr{B})$. In a sense of generality, the above method, specified by the decision function π, is one of many possible methods that operates on $\mathscr{B}$ and redefines its members as having required characteristics.

The two clustering algorithms discussed below basically exploit these two ideas of partitioning and repetitions. Depending on the exact algorithm, there might be some additional work; however, the basic principle always remains the same.

5.1 Adaptive K-Means Clustering for Big Data

We focus on big data with fixed dimension and huge number of individuals with velocity. Then, our discussion in Sect. 3 guides us to use a localized algorithm in order to cluster the dataset. We need to accommodate for noise in the data, growing number of clusters and other finer details for our algorithm to work. Rather than being explicit about the algorithm, we are going to present here a comprehensive sketch of what we call an *adaptive K-means* clustering for big data with fixed dimension. Tackling of noise is not considered here; we will discuss it in more details later.

Standard algorithmic practice for efficient algorithm in big data analytics is to work with pointers or indices to individual data points and process only a small number of indices at a time. This practice can be implemented quite easily for our proposed adaptive K-means algorithm. Let us consider a data stream of sizes $N_1, N_2, \dots, N_s, \dots$, i.e. on 'day' 5 we have data of size N_5. Suppose that at a time we can only accommodate around M data points for K-means clustering.

Partitioning—Partition each stream of the data so that each partition contains at most M data points. Here of course by partitioning the data we mean partitioning the indices; there is no need to consider the data yet. Let these partitions be $\{P_{11}, \dots, P_{1L_1}\}, \dots, \{P_{s1}, \dots, P_{sL_s}\}, \dots$, where $\{P_{s1}, \dots, P_{sL_s}\}$ is the partition of s-th data stream of size N_s.

Apply K-means algorithm on each of these partitions to get clusters $\{C_{sl}^1, \dots, C_{sl}^k\}$ from P_{sl}. Again it is only required to store the grouping of pointers of the clustering for further processing.

Combining—Start with the first set of clusters $\{C_{11}^1, \dots, C_{11}^k\}$. Consider the next set $\{C_{12}^1, \dots, C_{12}^k\}$. Use sequential comparing and combining operation similar to that discussed earlier in the section to identify distinct collection of clusters. For each $j = 1, 2, \dots, k$ check if C_{12}^j is "equal" to C_{11}^i for any i. This equality of two clusters can be conceived in many ways. One way is to check whether the two clusters are coming from the same probability distribution as a result of random sampling. One can also think whether the distance between these two clusters is so small that they can be taken as one single cluster. This checking of equality of two clusters can be defined by the user and implemented as a testing method accordingly. If there is one such i for a j then relabel the indices of C_{12}^j of being in C_{11}^i. If there is no such i then C_{12}^j is a new cluster.

We have mentioned in our earlier discussion that this operation is expensive both in terms of memory and computation. We can avoid this expense by applying the checking procedure of two clusters C and C' based on a representative subsample. Thus we can also follow our localized algorithm principle in order to save memory space and speed up the process.

The nicest thing of this very simple looking algorithm is that it allows a sequential processing. A data stream collected at high frequency can be very well adapted to this algorithm. The usual concern with choice of k is secondary here and can be quite easily accounted for using domain knowledge, e.g. if partitioning in the data stream is as per intervals of time then domain knowledge can let us know many similar groups of pattern we can expect in that interval of time. Consequently the combining method allows us to discover new clusters over time.

In this description of the algorithm we have deliberately ignored the possible noise in the data. To practically address this problem we need to put the K-means algorithm inside a wrapper that filters the noise along with extracting the clusters. An extensive discussion to address this issue is discussed in the next subsection where we have also proposed an algorithm that directly tackles the noise points and pick clusters at the same time.

Equality of two clusters can be checked in many ways. One can use any non-parametric multivariate test for equality of two distributions. However, usually they are not so powerful. We can use spanning tree and check the span of each cluster compared to the combined one. Then based on a certain criterion, we can decide upon whether two clusters are same. Another option is to use concentration ellipsoid and check whether it contains at least 90 % of points within the ellipsoid of the combined clustered data points. Some other adequate methods can also be thought of to settle this issue. The choice of test or logic behind combining two clusters might also depend on the objective that underlies the clustering of the data set.

5.2 Adaptation of Tight Clustering to Microarray Data

Microarray gene expression data is a type of data with high dimension but for considerably small number of individuals. We are interested in extracting prominent gene clusters which would shed light on finding co-regulation and specific pathways. This type of data is quite natural and common in big data analytics domain.

But as we have discussed earlier, there is a very challenging goal to be achieved while clustering such data. We are looking for clusters of genes which might be of tiny sizes (e.g. 10–50 genes out of hundreds of thousands) and they are very 'tightly' clustered. Thus most of the variables are present as unnecessary noise. Tseng and

Wong [18] proposed a detailed and carefully constructed algorithm, named 'tight clustering', to achieve this goal. Given a microarray data matrix of n genes, tight clustering algorithm runs K-means as a sub-algorithm based on a number of consecutive choices of k. For each of these different choices of k, it chooses the possible candidates for the tight-most cluster based on K-means algorithm applied on a number of subsampling of n genes. With these choices of tight-most cluster at hand, in the next step the algorithm picks the stable one, thus extracting the first stable tight set of genes that are likely to belong to a functional pathway or having some similarity. The second iteration for extracting the next set of similar gene patterns repeats the method on the remaining set of genes. Among several nice benefits of tight clustering a few are worth mentioning,

- Tight clustering method reduces the local minimum problem of K-means clustering to a certain extent.
- The algorithm always ensures stable clusters and extracted set of patterns that are ordered in accordance with their tightness.
- For the resulted clusters C_i's we have the relaxed constraint $\sum_{i=1}^{k} \# C_i \leq n$. As a result the choice of number of similar gene expression pattern becomes irrelevant.

The drawback of this algorithm is that it is computationally very demanding. To pick the tight-most and stable cluster it requires a clustering algorithm to be carried out a considerable number of times. While with a little intelligent implementation of the algorithm one can avoid falling into the trouble of memory problem, computational expense have to be paid. Our principle in Sect. 3 guides us to use a subsampling-based approach. Such a method would sample a portion of genes and cluster these subsamples and then output a cluster based on the clusters on the subsamples. But we have to keep in mind the target of extracting tightly clustered genes whose size might be very small. If we rely on a subsampling-based method, we would most certainly fail to reach that target. This is where we can take advantage of our partition-repetition principle.

5.2.1 Tight Clustering for Huge Number of Genes

Tight-most Cluster—Applying partition-repetition process, we get a bucket $\mathscr{B}$ where members of the bucket are candidates for the tight-most cluster from the whole data. To achieve this, on each partition of the data we apply tight clustering algorithm and collect the very first tight cluster and process this collection through the combining process. In order to obtain the tight-most cluster from the possible choices, i.e. members of $\mathscr{B}$, we require a measure of tightness of a set of data points. Suppose $\sigma(C)$ is a positive valued function defined on non-null subsets of X that tells us how disperse the set of data points C is. Then our desired cluster will be given by,

$$C^{(1)} := \arg \min_{C \in \mathscr{B}, C \neq \phi} \sigma(C). \tag{5}$$

The stability of this cluster is ensured as each output of the tight clustering method ensures a stable cluster.

Next Iteration—After we have obtained the tight-most cluster from the data, to get the next one in line, we remove all the data points of the obtained cluster from the whole data as well as from members of the bucket. Thus we are left with the pair $\left(X \setminus C^{(1)}, \mathscr{B} := \{B \setminus C^{(1)} | B \in \mathscr{B}\}\right)$.

Now the remaining data points are partitioned and tight-most cluster of each of these parts is put into the bucket $\mathscr{B}$. Notice that, unlike last time, we possibly already have a non-empty collections in the bucket. The operation $\mathbb{C}$ is carried out on the new bucket and just like the last time and the next in line tight-most cluster $C^{(2)}$ is obtained by comparing dispersion measure of the members of $\mathscr{B}$.

Continuing this way we extract k clusters, as $\left(C^{(1)}, C^{(2)}, \ldots, C^{(k)}\right)$ of sizes, respectively, $n_1, n_2, \ldots, n_k$ (say). The remaining data points, that do not belong to any cluster, are then identified as noise. The ability of filtering out the noise as a by product is a wonderful benefit of this iterative clustering algorithm.

In our discussion here, rather than providing the details about the algorithm and the analysis, we have focused on presenting the idea behind the implementation of the clustering that is adapted to big data. The adaptation is appropriate and justified mainly because partitioning and then the combining procedure ensures the quality of the cluster thus obtained.

6 Simulation Studies

In this section, our proposed algorithms are being demonstrated through two types of clustering as discussed in the last section. One is a version of K-means that is adapted to cluster big data. This method provides clusters from different segments/'days' of the whole data and attach cluster level to each data point. On the other hand, the other type of clustering that we discussed is an adaptation of tight clustering method. This method distributes a part of the whole dataset, in most cases a major part, to different clusters and then leave some data points as noise points. These noise points are interpreted as having entirely different properties that do not allow them to go into any of the clusters thus formed.

To evaluate these algorithms, we have done extensive simulations by generating different types of datasets with size of each data being very large. Our simulation shows that the algorithms hold good in principle nicely and can be applied to any real dataset that is huge in volume. In order to demonstrate the performance of adaptive K-means algorithm, we generate datasets of size 1000000 (1 Million) with varying dimension (p), i.e. for $p = 4, 6, 8, 10$, and 20. In each case we generate the data from five multivariate normal distributions so that it should have five clusters. Mean of each variable is taken to be $5(i-1)$ where i denotes the cluster level ($i = 1, \ldots, 5$). The standard deviation is always equal to 1 and pairwise correlation coefficient between two variables is taken to be 0.6. Thus if a cluster

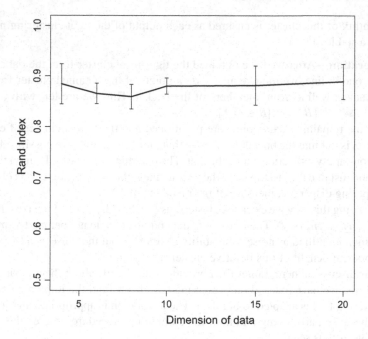

Fig. 1 Rand index (with 1 standard deviation margin for 10 runs) of adaptive K-means algorithm over varied dimension

level is 2, then each data point in that cluster is a random vector observation from $N_p(51_p, 0.6I_p + (1 - 0.6)11')$ where p is the dimension of the vector. The positive dependence between the variables ensures a small amount of overlapping between the clusters. To analyse the performance of the algorithm, we input the value of k as 5 and get the corresponding clusters after running our proposed adaptive K-means algorithm. In each case, since we know the true clusters, we can calculate Rand index to evaluate the performance of the algorithm. Higher the Rand index, better would be the performance of the clustering algorithm. In Fig. 1, we plot Rand index for the respective datasets for different dimensions. This is a good way to look how efficiently the algorithm works for varying dimension. The plot shows that the algorithm performs very satisfactorily in this set up. It is important to analyse the performance over varied dimension since random partitioning in the adaptive K-means algorithm could have had adverse effect on the quality of the clusters. The plot shows that this is not the case. Further, the run time of the algorithm is merely a couple of minutes on a non-parallelized implementation on a standard machine. The plotted figure is not chosen with any bias; very similar plots are seen for many other simulation scenarios with different k; k being varied over 'days' and different types of data clouds.

Generation of data that can be applied to see the efficiency of adaptive K-means algorithm is different than if we want to assess the performance of our modification of tight clustering approach. The main reason is that tight clustering method should be applied in some special situations where the general pattern of the data does not allow all points to go into one cluster or other. Rather, in such cases, there is a provision of a data point to remain alone in the data cloud and may be termed as noise. Microarray gene expression data is a nice example of one such situation. Keeping this in view, we have generated data that resemble microarray data. Based on a hierarchical lognormal distribution we generate this data [17] of size 100000 ($=10^5$) and for varying dimension as 5, 10, and 20. In each case we consider again only five clusters but vary percentage of noise as 5, 10, 15, and 20 %. Here again, since we know true clusters, we can calculate Rand index for each clustering and can check the efficiency of the algorithm. Moreover, plotting Rand index against percentage of noise reflects the effect of the amount of noise in the data. Also a plot against dimension would give us an idea how the performance of the algorithm changes over dimension. We run tight cluster algorithm specifically modified to adapt to big data for 10 times for a single data set and calculate the mean Rand index along with its standard deviation. We plot in Fig. 2, the mean Rand index against percentage of noise with error bars being equal to the standard deviation for each dimension. In Fig. 3, we plot Rand index against sample size across noise percentage. Finally in Fig. 4 we look at the performance as we increase dimension with varying noise level. These graphs clearly explores the performance of our algorithm that leaves a portion of data as noise.

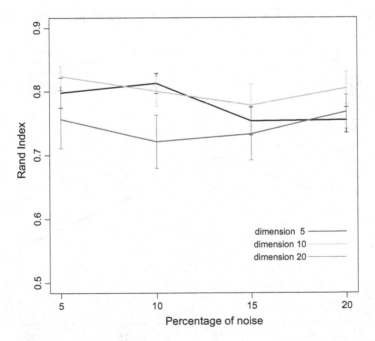

Fig. 2 Rand Index (with 1 standard deviation margin for 10 runs) of proposed tight clustering algorithm across dimension and varied noise level

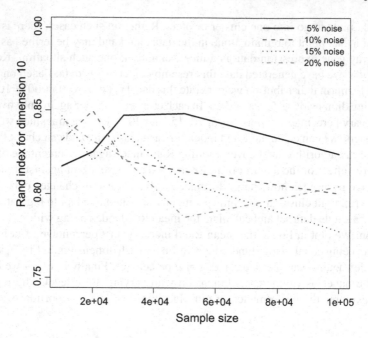

Fig. 3 Rand Index of proposed tight clustering algorithm across sample sizes and varied noise level

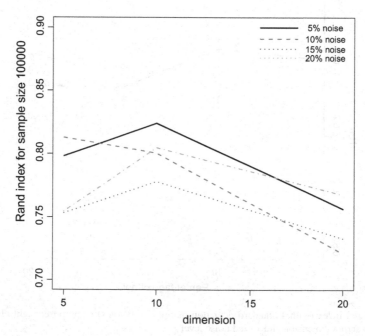

Fig. 4 Rand Index of proposed tight clustering algorithm across dimension and varied noise level

7 Concluding Remarks

Huge volume, diverse variety and high velocity sometimes make the standard statistical tools challenging in the domain of big data analytics. Clustering is definitely one such area where some modification of the existing algorithms as well as the adaptation and implementation of techniques to big data is required. Clustering algorithms demand computational efficiency along with scientifically reasonable arguments that ensure picking good quality clusters from the data cloud. Naturally this faces considerable challenges when the data volume is too large and grow with high velocity. Current state of the art does not discuss this issue in a rigorous way and is little scattered. Existing literature mainly evaluates the existing clustering algorithms to its adaptability to big data domain. But focussed research is needed towards resolving this challenge.

To this end, we have proposed two algorithms that are adapted specifically to tackle the big data issues in order to find good quality clusters that are amenable to further statistical analysis. These two algorithms are suited for two separate big data scenarios. First one of them is an adaptive version of K-means algorithm that is proposed for data with very high velocity but fixed dimension. The later is a scalable version of tight clustering algorithm that is proposed for high dimensional gene expression data or medical imaging data or any other data having this feature.

Adaptive K-means algorithm is very fast and cluster data points efficiently. This method as we proposed can be used in a sequential manner and combined with the previously obtained clusters as and when required. It can be modified and applied to a portion of data instead of the whole data and can easily ignore data beyond a certain time in the past if required by the situation. Naturally this flexibility is quite appealing to tackle the problems due high velocity of data flow. As apparent from Rand index values in our simulation study, our adaptation also ensures good quality clusters, even after partition of relevant or available portion of dataset, that is usually very large.

Tight clustering method specifically adapted to big data clustering in presence of noise is an interesting one to identify scatter points even with considerable velocity of data coming into the domain of study. Noise points may arise due to many reasons, an important one being the diverse variety of data. Since partitioning is an inherent feature of our proposed adaptation of tight cluster algorithm, it automatically takes care of the 'variety' characteristic of the data. Although little more computation intensive, this algorithm is able to tackle the volume of big data in an efficient manner as transpired from Rand index values in our simulation study and is quite effective over dimensions and increasing percentage of noise.

Idea of partitioning is instrumental in tackling the large volume of data, but combination wrapper inside our algorithm ensures the quality and quantity of clusters thus obtained. Our objective is not just to apply clustering algorithm to a partition of a dataset, but the main issue is to find clusters that can be obtained if there is hypothetically huge space to consider the whole data together. Hence combining different similar type clusters is the key to produce the end product and hence should be done efficiently and quickly without losing mathematical and statistical rigour. Our proposed wrapper is able to find reasonable number of clusters with very good quality making a sense of the data cloud as a whole.

We have written R code to implement these algorithms. They can be easily made parallel if one wants to do the computation in a computing cluster consisting of many nodes or in different machines. Even the data storage can be done in separate boxes or memories so as to apply the algorithm. Although we describe here only two methods, the basic idea can be applied to any standard clustering method thus enabling us to produce good quality clusters from a real big data. Many functions that are wrapped inside the algorithm can also be chosen by the user depending on the real need of the problem.

As a future research direction we want to make a note of the fact that, in terms of clustering algorithms an open field remains where velocity brings with it increasing number of features for individuals. In Sect. 3 we have provided a peek into possible algorithmic principles for such data. A practical method for clustering such data would be beneficial to large consumer facing industries.

References

1. Aggarwal CC, Reddy CK (2013) Data clustering: algorithms and applications, 1st edn. Chapman & Hall/CRC
2. Ankerst M, Breunig MM, Kriegel HP, Sander J (1999) Optics: ordering points to identify the clustering structure. SIGMOD Rec 28(2):49–60
3. Berkhin P (2006) A survey of clustering data mining techniques. In: Kogan J, Nicholas C, Teboulle M (eds) Grouping multidimensional data. Springer, Berlin, pp 25–71
4. Ester M, Kriegel HP, Sander J, Xu X (1996) A density-based algorithm for discovering clusters in large spatial databases with noise. AAAI Press, pp 226–231
5. He Y, Tan H, Luo W, Feng S, Fan J (2014) MR-DBSCAN: a scalable mapreduce-based dbscan algorithm for heavily skewed data. Front Comput Sci 8(1):83–99
6. Hinneburg A, Hinneburg E, Keim DA (1998) An efficient approach to clustering in large multimedia databases with noise. AAAI Press, pp 58–65
7. Hinneburg A, Keim DA (2003) A general approach to clustering in large databases with noise. Knowl Inf Syst 5(4):387–415
8. Januzaj E, Kriegel HP, Pfeifle M (2004) DBDC: Density based distributed clustering. In: Bertino E, Christodoulakis S, Plexousakis D, Christophides V, Koubarakis M, Bhm K, Ferrari E (eds) Advances in database technology—EDBT 2004. Lecture notes in computer science, vol 2992. Springer, Berlin, pp 88–105
9. Karypis G, Kumar V (1999) Parallel multilevel k-way partitioning for irregular graphs. SIAM Rev 41(2):278–300
10. Katal A, Wazid M, Goudar RH (2013) Big data: Issues, challenges, tools and good practices. In: 2013 Sixth International conference on contemporary computing (IC3), pp 404–409
11. Kaufman L, Rousseeuw PJ (2005) Finding groups in data: an introduction to cluster analysis. Wiley Interscience
12. Kim Y, Shim K, Kim MS, Lee JS (2014) DBCURE-MR: an efficient density-based clustering algorithm for large data using mapreduce. Inf Syst 42:15–35
13. Kleinberg JM (2003) An impossibility theorem for clustering. In: Becker S, Thrun S, Obermayer K (eds) Advances in neural information processing systems, vol 15. MIT Press, pp 463–470

14. Kogan J, Nicholas C, Teboulle M (2006) A survey of clustering data mining techniques. Springer, Berlin, pp 25–71
15. Ng RT, Han J (2002) CLARANS: A method for clustering objects for spatial data mining. IEEE Trans Knowl Data Eng 14(5):1003–1016
16. Shirkhorshidi A, Aghabozorgi S, Wah T, Herawan T (2014) Big data clustering: a review. In: Murgante B, Misra S, Rocha A, Torre C, Rocha J, Falco M, Taniar D, Apduhan B, Gervasi O (eds) Computational science and its applications ICCSA 2014. Lecture notes in computer science, vol 8583. Springer International Publishing, pp 707–720
17. Thalamuthu A, Mukhopadhyay I, Zheng X, Tseng GC (2006) Evaluation and comparison of gene clustering methods in microarray analysis. Bioinformatics 22(19):2405–2412
18. Tseng GC, Wong WH (2005) Tight clustering: a resampling-based approach for identifying stable and tight patterns in data. Biometrics 61(1):10–16
19. Wang S, Fan J, Fang M, Yuan H (2014) Hgcudf: hierarchical grid clustering using data field. Chin J Electron 23(1):37–42
20. Zhang T, Ramakrishnan R, Livny M (1996) BIRCH: an efficient data clustering method for very large databases. SIGMOD Rec 25(2):103–114
21. Zhao W, Ma H, He Q (2009) Parallel k-means clustering based on mapreduce. In: Jaatun M, Zhao G, Rong C (eds) Cloud computing, vol 5931., Lecture notes in computer science Springer, Berlin, pp 674–679

Online Graph Partitioning with an Affine Message Combining Cost Function

Xiang Chen and Jun Huan

Abstract Graph partitioning is a key step in developing scalable data mining algorithms on massive graph data such as web graphs and social networks. Graph partitioning is often formalized as an optimization problem where we assign graph vertices to computing nodes with the objection to both minimize the communication cost between computing nodes and to balance the load of computing nodes. Such optimization was specified using a cost function to measure the quality of graph partition. Current graph systems such as Pregel, Graphlab take graph cut, i.e. counting the number of edges that cross different partitions, as the cost function of graph partition. We argue that graph cut ignores many characteristics of modern computing cluster and to develop better graph partitioning algorithm we should revise the cost function. In particular we believe that message combing, a new technique that was recently developed in order to minimize communication of computing nodes, should be considered in designing new cost functions for graph partitioning. In this paper, we propose a new cost function for graph partitioning which considers message combining. In this new cost function, we consider communication cost from three different sources: (1) two computing nodes establish a message channel between them; (2) a process creates a message utilize the channel and (3) the length of the message. Based on this cost function, we develop several heuristics for large graph partitioning. We have performed comprehensive experiments using real-world graphs. Our results demonstrate that our algorithms yield significant performance improvements over state-of-the-art approaches. The new cost function developed in this paper should help design new graph partition algorithms for better graph system performance.

Keywords Graph partition · Heuristic · Cost function · Online algorithm

X. Chen · J. Huan (✉)
Information and Telecommunication Technology Center,
The University of Kansas, Lawrence 66047, USA
e-mail: jhuan@ittc.ku.edu

© Springer India 2016
S. Pyne et al. (eds.), *Big Data Analytics*, DOI 10.1007/978-81-322-3628-3_6

95

1 Introduction

Graph-structured data are playing a central role in data mining and machine learning. Such data are widely used in application domains where linkages among objects are as important as the objects. Examples include the web, social networks, semantic web, recommender systems and biological networks among others.

Large-scale graph data ask for scalable systems that can scale to graphs with millions or even billions of nodes. There are several systems supporting intensive computations on input graph data. Examples include *Pregel* [1], *Giraph* [2] and *Spark* [3]. There are also systems support iterative machine learning computations such as approximate Bayesian inference in systems such as *Graphlab* [4], or for efficiently resolving queries in large-scale databases *Neo4j* and *Trinity* [5].

In designing systems supporting large-scale graph mining and learning systems a key question is graph partitioning. In graph partitioning we map vertices of a graph to computing nodes in a cluster so that the overall computation efficiency is optimized. In such mapping we considered two critical factors: communication cost and load balancing. Since we map vertices to different nodes, there is a natural problem of inter-node communication cost where we send a message from a node to another one for information exchange or computing synchronization. Since computing on graphs is typically iterative and different vertices may have different amounts of computing, in load balancing, we would like to have the state where each computing node has similar work load. As studied before graph partitioning often affects the computational performance of any graph mining and learning systems dramatically [6]. Many heuristic and approximation algorithms were proposed to solve the related optimization problems in graph partitioning [7–9].

Reducing communication cost is one of the most important problems in graph partitioning and how to devise the cost function is the key issue. Several popular graph computing algorithms, including *Pregel* and *Graphlab*, take graph cut, i.e. counting the number of edges that cross different partitions, as the cost function of graph partition. Graph cut is a simplified abstraction of the communication cost and it ignores many characteristics of modern computing clusters. For example, the communication cost on creating a new message channel between the different partitions is far greater than exchanging a short message between the two machines.

To design better graph partitioning algorithms, we argue that different communication patterns should be considered. Formally, we proposed a new cost function for graph partitioning including message combining. In the new cost function, we consider communication cost from three different sources: (1) the cost that two machines establish a message channel between them; (2) the cost that a process creates a message utilizing the channel and (3) the cost that a process sending a message through the established communication channel. Based on this cost function, we developed several heuristics for large graph partitioning. We call this function affine cost function since we use a weighted linear combination of the three factors. Comprehensive empirical experiments on real-world graphs demonstrate that our algorithms yield significant performance improvements over state-of-the-art approaches.

The rest of the paper is structured as follows. In Sect. 2 we survey current graph partitioning algorithms, grouped by the heuristic that were used. In Sect. 3 we introduce the background of current cost function and three heuristic algorithms. In Sect. 4 we define affine graph partitioning cost function. In the same section we introduce two online heuristic algorithms solving the related optimization problems. We present the empirical evaluation of the new algorithm together with those from state-of-the-art heuristic algorithms in Sect. 6. Conclusion and discussion are described in Sect. 7.

2 Related Work

Graph partitioning is often an NP-hard problem and solutions are generally derived using approximation and heuristic algorithms.

Traditional approximate solution focuses on the (k, v)-balanced partitioning problem. In the (k, v)-balanced partitioning problem we divide the graph of n vertices into k partitions; meanwhile, the number of the vertices of each partition is not greater than $v * n/k$. In the special case of $k = 2$, the problem is called the Minimum Bisection problem. Several approximation algorithms exist for the minimum bisection problem [7, 9]. Some of those algorithms can be extended to larger values of k. The algorithm by Saran and Vazirani gives an upper bound of $n * (k - 1)/k$ for the $(k, 1)$-balanced partitioning problem. Note however that this algorithm requires a running time that is exponential in k. A critical limitation of approximation graph partitioning algorithms is that they often run in quadratic time. For graphs with millions or billions of nodes, linear or sublinear running time is preferred and quadratic processing time is prohibitive.

Heuristic algorithms are widely used in graph partitioning as well. We categorize such problems into two groups: offline graph partitioning and online graph partition.

A common offline approach is to construct a balanced k-way cut in which sub-graphs are balanced over machines and communications between machines are minimized. Offline methods, such as *Spectral Clustering* [10], *Metis* [11] and *K-partitioning* [12], collect full graph information to perform offline partitioning. For example, widely used algorithm *Metis* uses a heuristic algorithm that iteratively reduces graph sizes with multiple stages. Once we reach a workable size, we start the partitioning phase from the smallest graph and refine the partition from smaller graphs to larger graphs. Metis usually achieves good cuts but exhibit poor scaling due to high computation and memory costs [13]. Especially, these algorithms perform poorly on power-law graphs and are difficult to parallelize due to frequent coordination of global graph information.

In online graph partitioning vertices (or edges) arrive as a steam. We place each vertex (edge) using a *"on-the-fly"* approach in order to minimize the overhead of computing. Two commonly used classes of such algorithms are those based on the hash or greedy heuristic.

Hash or random partition is the simplest method. It supports both vertex and edge partitions. Random partition of vertices is first initiated and is popular for many graph processing platforms such as *Pregel* [14], and related algorithms such as *Apache Giraph*, *PEGASUS* [15] and *GraphLab*. Vertex random partition assigns graph vertices to one of the k machines uniformly at random. This heuristic method is efficient for balancing the number of vertices over different clusters but ignores entirely the graph structure and may not perform well on minimizing the communication cost. To cope with the poor performance of vertex random partition on real-world power-law graphs, such as social networks and the web, *Powergraph* [16] proposed random edge partition which assigns each edge rather vertex to one of the k partitions uniformly at random and employ a "master-mirror" technique to keep the consistent of the nodes "split" in different clusters. Theoretical and empirical studies have demonstrated the advantage of edge partition over vertex partition [16, 17].

Greedy algorithms improve the random algorithm by introducing heuristics to vertex or edge assignments. The recent greedy algorithms introduced and discussed by Stanton and Kliot [18], as well as the latest Fennel method [19], usually consider some greedy rules in graph partition. For example, the *DG* (*Deterministic Greedy*) and *LDG* (*Linear Weighted Deterministic Greedy*) [18] algorithms usually assign a vertex (edge) to the partition that holds the largest number of its neighbours.

From the cost function point of view current approximation algorithm or heuristic algorithm only consider the number of edge cut in computing the communication or message cost. Such simplified abstract has the potential to be improved by incorporating the different kinds of message costs which are closer to real cluster systems.

3   Background

In this section, we introduce notations that we use throughout this paper.

A directed graph is a tuple where $G = (V, E)$ with the set of vertices V and the set of directed edges $E \subset V \times V$. $n = |V|$ and $m = |E|$ are number of vertices and the number of directed edges, respectively. Let $N_{in}(u) = \{v \in V, (v, u) \in E\}$ and $N_{out}(u) = \{v \in V, (u, v) \in E\}$ denote the set of vertices that have edges, respectively, coming into and leaving away from $u \in V$. $d_{in}(u)$ and $d_{out}(u)$ are the in-degree and out-degree of vertex $u \in V$, i.e. $d_{in}(u) = |N_{in}(u)|$ and $d_{out}(u) = |N_{out}(u)|$. Note that $\sum_{u \in V} d_{in}(u) = \sum_{u \in V} d_{out}(u)$. Let S_i be the set of vertices in partition i ($S_i \subseteq V$). Let $N(v)$ be the set of vertices that are neighbours of vertex v. That is to say the vertex in $N(v)$ is incident to or from vertex v. In our context, a partition is the set of vertices that are assigned to the same machine (computing node) in a computer cluster. In this sense, the two terms, a partition and a machine, are interchangeable.

We use a $|V| \times k$ *binary* matrix X to represent a partition schema of a graph with $n = |V|$ vertices into k machines. We define $X_{u, i} = 1$ if vertex $u \in V$ is assigned to a machine $i \in [1 \ldots k]$, and $X_{u,i} = 0$ otherwise. Each vertex is assigned to exactly one machine and for any valid graph partitioning schema X we have $\sum_{i=1}^{k} X_{u,i} = 1$ for all $u \in V(G)$.

Each graph partitioning problem is characterized by a communication cost (denoted by C) which has to be minimized under a condition of balanced loads of the clusters. If we denote $L_i(X) = \sum_{u \in V} X_{ui}$ the load of machines $i \in [1 \ldots k]$, i.e. the set of graph vertices that are assigned to the machine i for a schema X, the graph partitioning problem can be written as

$$
\begin{cases}
\text{Minimize: } C(X) \\
\text{Subject to: } \text{Max}_{i \in [1..k]} L_i(X) \leq \lambda \frac{\sum_{i \in [1..k]} L_i(X)}{k}
\end{cases} \tag{1}
$$

Here, v is an off-balancing coefficient.

A way to define the cost function C used in previous equation is to count the number of edges cross different machines. For a cluster with k machine, this cost function is formulated by

$$
C(X) = \sum_{i=1}^{k} \sum_{(u,v) \in E} X_{ui}(1 - X_{vi}) \tag{2}
$$

This cost function is used by many current heuristic algorithms. There are three widely used heuristic online algorithms. They are compared with our algorithms in Sect. 5.

- *Random/Hash Partition:*
  S_i chose uniformly by random. Assign v to partition i, we use $i = (random number \bmod k) + 1$;
- *Deterministic Greedy (DG):*
  Maximum $|N(v) \cap S_i|$
- *Linear Weighted Deterministic Greedy (LDG): Maximum* $|N(v) \cap S_i|\left(1 - |S_i|/\left(\frac{n}{k}\right)\right)$

The random partition heuristic is the simplest method to partition the graph and is widely used. *Deterministic Greedy algorithm (DG)* considers reducing the edge cut by assigning the vertex to the partition including most number of v's neighbours. Based on *DG*, *Linear Deterministic Weighted Greedy algorithm (LDG)* uses an item $(1 - |Si|/(n/k))$ to balance the partition.

3.1 Cost Function with Message Combining

In the previous section, the communication cost is simply the number of edges that cross different partitions. This may not reflect a realistic situation. Pregel proposes a

Fig. 1 Vertex partition with combining

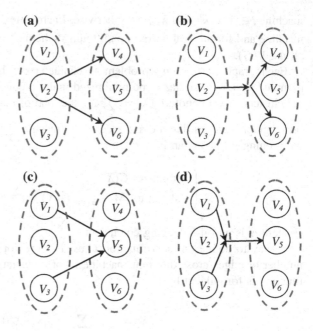

method called message combining and allows users to combine messages that are sent from a partition to the same vertex v in another partition. These messages are first combined into a single message and then sent to the vertex v in another partition. Similarly, the message sent from one vertex v in a partition to the different vertexes in another partition can also be combined into a single message (Fig. 1).

Formally, the cost function (C^{mc}) with message combining is [17]

$$C^{mc}(X) = \sum_{u \in V} \sum_{i=1}^{k} X_{ui} \sum_{j \neq i} \left[\left(1 - \prod_{v \in N_{in(u)}} (1 - X_{vj}) \right) + \left(1 - \prod_{w \in N_{out(u)}} (1 - X_{wj}) \right) \right]$$

(3)

Current heuristic graph partitioning algorithms do not consider combining in graph partitioning. The common method to handle combining is to first partition the graph and then use combining strategy to reduce the message or edge cut.

4   Methodology

4.1   Cost Function with Affine Message Combining

In this section we consider a more realistic cost function that can be applied in modern cluster computing environment.

We notice that in current clusters sending a message from one partition to another partition may have different costs depending on several factors. Creating a new message channel between two different partitions that have not message channel between them will incur overhead larger than sending a new message between two partitions that have already a message channel. In addition, if two messages between two partitions can be combined to one message, the communication overhead incurred by such a combined message is less than that of two messages that cannot be combined. Based on this consideration we argue that different types of graph cut and combining issue should be considered in designing optimal cost functions for graph partitioning.

We use an affine function to formalize a cost function that we believe better characterize communication cost than graph cut, as shown below:

$$C(x) = w_1 \times n_1 + w_2 \times n_2 + w_3 \times n_3 \tag{4}$$

In this cost function w_1 denotes the cost of creating a new message channel between two different partitions. w_2 denotes the cost of creating a message utilizing the channel and w_3 is the cost of sending a message with one unit length. Here the length of the message means the number of messages that have been combined to one message on condition the combining method. In real clusters w_1 is far greater than w_2 and w_2 is greater than w_3. n_1, n_2 and n_3 are numbers of those three different messages (edges) in the graph. Figure 2 is an example to illustrate the cost of different types of messages.

We illustrate the new cost function in Fig. 2. There are message channels between partition M_1 and M_2 as well as M_2 and M_3 so n_1 is 2; there exists two messages utilized the message channel between partition M_1 and M_2: one is the combined message (V_1V_5 combined with V_2V_5) and the other is a single message (V_6V_3). Their message lengths are two and one separately. Therefore, n_2 is 2 and n_3 is 3 for messages between partition M_1 and M_2. Similarly, there exists one message ($n_2 = 1$) utilized, the message channel between partition M_2 and M_3, and the length of such message is three (V_4V_7, V_4V_8 and V_4V_9 are combined to one message. Then here $n_3 = 3$.). The overall cost of the graph partitioning illustrated in Fig. 2 is : $2 \times w_1 + 3 \times w_2 + 6 \times w_3$.

Fig. 2 The different messages between partitions

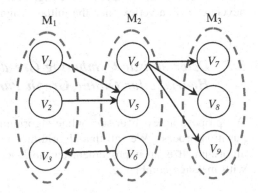

According to the definition of n_1, n_2 and n_3 in (4) they can be formally formulized as

$$n_1 = \sum_{(i,j) \in [1..k]} \sum_{(u,v) \in E} \left(1 - \prod (1 - X_{ui}X_{vj})\right)$$

$$n_2 = \sum_{(i,j) \in [1..k]} \sum_{u,v \in V} X_{ui}\left(1 - \prod (1 - X_{vj})\right)$$

$$n_3 = \sum_{(i,j) \in [1..k]} \sum_{(u,v) \in E} X_{ui}(1 - X_{vj})$$

The formal formula of the cost function with affine message combining is

$$C^{\text{affine}}(X) = w_1 \times \sum_{(i,j) \in 1}^{k} \sum_{(u,v) \in E} \left(1 - \prod (1 - X_{ui}X_{vj})\right) + w_2$$

$$\times \sum_{(i,j) \in [1..k]}^{k} \sum_{u,v \in V} X_{ui}\left(1 - \prod (1 - X_{vj})\right) + w_3 \times \sum_{(i,j) \in [1..k]}^{k} \sum_{(u,v) \in E} X_{ui}(1 - X_{vj})$$

$$(5)$$

5 Online Heuristics for Cost Function with Message Combining

In this section, we present two online heuristics for graph partition. The first algorithm uses the cost function with message combining (Eq. 3) and the second algorithm use the cost function with affine message combining (Eq. 5).

The online graph model has been described in previous work [18, 19]. We briefly outline the setup here. In this algorithm we have a computing cluster of k partitions, each with memory capacity λ and the total capacity, λk, is large enough to include the whole graph. The input graph may be either directed or undirected. Undirected graph can be taken as directed graph where each edge is bi-directional. As vertices arrive, our algorithm assigns the vertex to one of the k partitions. We never relocate a vertex once the initial assignment is made.

5.1 Deterministic Combining Greedy Algorithm: A Message Heuristic for Online Graph Partitioning

We propose a new heuristic online algorithm that considers the combining in the partition process. We call this algorithm *Combining Deterministic Greedy* (CDG) algorithm. It is an improvement on the *LDG* heuristic based on the cost function with message combining.

To clarify the score function used in CDG algorithm, we first define some formulation:

$$N(v) = \{w|wv \in E \text{ or } vw \in E, w \in V\}$$
$$N(S_i) = \{v|\exists w \in S_i, vw \in E \text{ or } wv \in E\}$$
$$\Phi(v, S_i) = |N(v) \cap S_i|$$
$$\theta(v, S_i) = |N(v) \cap N(S_i)|$$
$$\mu(S_i) = (1 - |S_i|/(n/k))$$

Here $N(v)$ is defined as the *neighbours* of vertex v. In another word, it is the set of the vertices that are incident to or from vertex v.

Similarly, $N(S_i)$ is defined as the *neighbours* of partition S_i. In another word, it is the set of vertices that are incident to or from any vertex in partition S_i.

The item $|N(v) \cap S_i|$ is also used by previous algorithms, such as DG and LDG. It computes the number intersection of vertex v's neighbours with the vertices in partition S_i. This item evaluates the message reduction by allocating vertex v with its neighbours in the same partition.

The item $|N(v) \cap N(S_i)|$ is the number of intersection of vertex v's neighbours with the neighbours of partition S_i. This item evaluates the message reduction by combining the message incident to or from partition S_i.

$\mu(S_i)$ is used to balance the load of partitions.

The score function used in *CDG* algorithm is

$$S_{cdg} = [\Phi(v, S_i) + \alpha \times \theta(v, S_i)] \times \mu(S_i) \qquad (6)$$

Since the S_{CDG} is already has the item $\mu(S_i)$ to balance the partition. The heuristic of CDG algorithm is simplified:

$$\text{Maximum: } S_{cdg}$$

This demonstrates that *CDG* not only tries to allocate the vertex v with its neighbours in the same partition but also tries to utilize the message combining and allocate v to the partition that more combining is possible. The parameter α is the weight for combining item in the process of graph partition.

We use a simple case (Fig. 3) to explain the graph partitioning process in *CDG*:

When a new vertex V_6 arrives, we need to consider three possible assigning, that is, case1 for assigning V_6 to partition S_1, case2 for assigning V_6 to partition S_2 and case3 for assigning V_6 to partition S_3.

The previous heuristics (such as *DG* and *LDG*) will give case2 and case3 the same score because both S_2 and S_3 include the same number of neighbours of V_6. However, *CDG* will give case 2 higher score because there exits another V_6's neighbour—V_1 in partition S_1 and at the same time V_1 is also the neighbour of the vertex V_2 in the partition S_2 ($V_1 \in N(S_i)$). The edge $V_1 V_2$ and $V_1 V_6$ has the same

Fig. 3 Assign a new vertex
using online heuristic

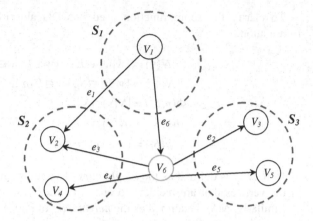

direction, which means if we assign V_6 to partition S_2 the edge V_1V_6 and edge V_1V_2
can be combined, thus avoid creating a new cut.

Formally, the CDG algorithm is presented below.

CDG algorithm:
[1]Initialize Step:
1: for all the vertex v in graph do assign(v) = 0
[2]Greedy Step:
1: for each v_t in V do
2:   for each partition S_i do
3:     (consider each $w \in N(v_t)$ that has already been allocated)
4:     computing $\Phi(v_t, S_i)$;
5:     computing $\Theta(v_t, S_i)$;
6:     get $score_i$ by score function (6)
7:   end for
8:   $m = \arg \max \{ score_i \}, i \in [1, k]$
9:   add v_t to partition S_m and set assign(v_t) = m
10: end for

5.2 Affine Combining Deterministic Greedy Algorithm: A Online Heuristic with Affine Message Combining

In this section we present an online heuristic algorithm *Affine Combining Deterministic Greedy* (*ADG*) algorithm for graph partitioning based on the cost function
with affine message combining (Eq. 5).

For a new arriving vertex, *ADG* computes and compares the scores (cost) on
condition that assigning this vertex to every partition and select the partition with
the optimal (minimum) score to assign the vertex. This procedure will be repeated
until all the vertices have been assigned to the partitions. The score function used by

ADG is based on the cost function (Eq. 5). Besides, a balancing item $(1 - |S_i|/(n/k))$ is incorporated to the score function to guarantee balancing. In addition, the item $|N(v) \cap N(S_i)|$ is also incorporated since it reflects the better partition from another angle.

The heuristic score function used in *ADG* is

$$S_{adg} = [C^{affine} / (\alpha \times \theta(v, S_i) + 1)] \times 1 / \mu(S_i) \qquad (7)$$

Since the S_{CDG} is already has the item $\mu(S_i)$ to balance the partition, the heuristic of ADG algorithm is simplified:

Minimum: S_{adg}

Here cost C^{affine} is computed by Eq. 5. It denotes the message overhead incurred by adding a new vertex to the partition. Parameter α is a weight factor.

Specifically, *ADG* includes the following steps:

In the initial step, each of the vertices v $(v \in V)$ in Graph is set as not assigned $(assign(v) = 0)$ and there is no message channel between any pair of the partitions $(C_{ij} = 0, i, j \in [1..k]$ and k *is the number of partitions*).

In step of adding a new vertex v_t, we first get the information of v_t, including the neighbours of v_t $(N(v_t))$ and edge information that indent to or from v_t. Then we evaluate each candidate partition and find the optimal one according to the score by assigning v_t to the partition.

In the process of evaluating each candidate partition, for example, evaluating the candidate partition i, we deal with five different cases:

(1) vertex w, the neighbour of v_t is not assigned before; we just ignore w.

(2) w is already assigned to the same partition i. In this case, if we assign v_t to partition i, there will no message overhead incurred.

(3) w is already assigned to a partition that is different with partition i. At the same time there is no message channel between those two partitions. We need to create a new message channel between those two partitions.

(4) w is already assigned to a partition that is different with partition i. There already exists a message channel between those two partitions. If the edge $v_t w$ can be combined with a message in this message channel (according to the combining definition, they have the same source vertex or end vertex) we only need adding the length of this message.

Otherwise (5), we need create a new message to utilize the message channel between partition i and p.

The *ADG* algorithm is described below:

ADG algorithm:

[1]Initialize Step:
  1: for all the vertex v in graph do assign(v) = 0
[2]Greedy Step:
  1: for each v_t in V do
  2:    for each partition S_i do
  3:       (consider each $w \in N(v_t)$ that has already been allocated)
  4:       if w is also allocated in partition S_i
  5:          computing $\Theta(v, S_i)$
  6:       else
  7:          computing C^{affine} by equation 5
  8:       compute the $score_i$ by score function (7)
  9:    end for
  10:   $m = \arg max \{ score_i \}, i \in [1, k]$
  11:   add v_t to partition S_m and set assign(v_t) = m
  12: end for

6 Experiment

We present the experimental results of our algorithms in comparison with the *Random, DG, LDG* and *Metis* algorithms. It should be noted that *Metis* is the offline algorithm and the others are online algorithms. Here we compare the results with *Metis* since it is often used as an ideal reference to evaluate the performance of online partitioning algorithms.

Below Sect. 6.1 describes the experimental setup. Sections 6.2 and 6.3 present our results using our online heuristic algorithms that employ two different cost functions, respectively. The efficiency of *CDG* and *ADG* is measured experimentally in Sect. 6.4. Finally, we further explore the performance of those algorithms in different stream orders.

6.1 Experimental Setup

The real-world graphs data used in our experiments are shown in Table 1. They are directed graphs and publicly available on the Web [20]. Some ineffective edges that include the vertex number out of the range of the definition on the dataset itself are removed, if any (Table 1).

- **Evaluation Criteria**

The performance of an algorithm is measured using two criteria from the resulting partitions:

(1) The first criteria is widely used by state-of-the-art algorithms to measure the performance of graph partition: For a fixed partition X, we use the measures of

Table 1 Datasets used in our experiments

Graph Data	Nodes	Edges	Description
p2p-Gnutella04	10876	39994	P2P
P2p-Gnutella05	8846	31839	P2P
P2p-Gnutella06	8717	31525	P2P
P2p-Gnutella08	6301	20777	P2P
P2p-Gnutella09	8114	26013	P2P
P2p-Gnutella24	26518	65369	P2P
P2p-Gnutella25	22687	54705	P2P
P2p-Gnutella30	36682	88328	P2P
P2p-Gnutella31	62586	147892	P2P
Slashdot0811	77360	905468	Social
Slashdot0902	82168	948464	Social
Soc-Epinions1	75879	508837	Epinions.com
Soc-sign-Slash	77357	516575	signed social network
Wiki-Vote	7115	103689	Wiki network
Amazon0302	262111	1234877	Co-purchasing
Amzaon0601	403394	3387388	Co-purchasing

the fraction of edges cut γ and *the normalized maximum load* Δ. In addition, we use the measures of the *fraction of combined edges cut* $\gamma_{combine}$ to measure the performance of each algorithm with message combing (That is to say, we take edges cut that can be combined as one edge cut). They are defined as

$$\gamma = \frac{\text{Edge Cuts by } X}{|E|} \tag{8}$$

$$\gamma_{combine} = \frac{\text{Combined Edge Cuts by } X}{|E|} \tag{9}$$

$$\Delta = \frac{\text{Maximum Load}}{n/k} \tag{10}$$

Notice that $n \geq \gamma \geq n/k$ since the maximum load of a partition is at most n and there always exists at least one partition with at least n/k vertices.

(2) The other criterion uses cost but the edge cuts to evaluate the relative improvement of competed algorithms over the random method. The cost of the resulting partition is computed by our new cost functions with combining. We define *Cost* $_m$ the cost of the partition by method m and take the resulting

partition of random as our base reference to evaluate other four methods. The *relative cost ratio* ρ is defined as

$$\rho = \frac{\text{Cost}_m}{\text{Cost}_{\text{random}}} \qquad (11)$$

Here ρ ranges from 0 (there is not any message cost in resulting partition) to 1 (the resulting partition has the same message cost as that of random method). From this definition the smaller ρ indicates the less cost compared to that of random method and shows the better performance.

6.2 Partition Results Using CDG Heuristic

To evaluate the performance of our heuristic algorithm (*CDG*) based on cost function with message combining, we compare it with four state-of-the-art heuristics: *Random*, *DG*, *LDG* and *Metis* on condition using the same message combining cost function described in Sect. 3.1.

To get the resulting partition with message combining of four competing algorithms, we first use those algorithms to get the resulting partition and then apply message combining to their resulting partitions. Specifically, to evaluate the effecting of message combing, we define γ as the edge cut fraction (same with the definition of γ) and γ_c as the combined edge cut fraction (same with the definition of γ_{combine}).

As shown in Tables 2 and 3 the performance (γ_c) of *CDG* is close to Metis and outperforms Random and *LDG* algorithm. In case that the number of partitions exceeds 64 *DG* shows better results on γ_c than *CDG* but it performs very poor in load balance (Δ). The other four algorithms perform well on this issue.

Table 4 presents detailed results including the edge cut fraction γ, combining edge cut fraction γ_c and the load balance Δ on a dataset *Slashdot0811*. *CDG* shows better performance than Random, LDG on all different numbers of partitions. It even shows better performance on combining edge cut fraction when the number of partitions is larger than 16. DG still shows the best results on both γ and γ_c but the worst load balance.

Table 2 Average combining edge cut fraction γ_c on different numbers of partitions

K	Average combining edge cut fraction γ_c							
	2	4	8	16	32	64	128	256
Random	4.6	14.7	30.2	46.6	60.1	72.3	81.3	88
DG	4.0	11.0	19.3	26.5	31.4	34.6	36.7	37.7
LDG	4.2	11.6	20.8	29.5	36.8	42.9	48	52.7
CDG	3.7	9.8	16.7	24.1	29.6	34.8	41.5	45.9
Metis	3.2	8.7	15.1	21.3	27.0	32.5	37.1	43.2

Table 3 Average load balance Δ on different numbers of partitions

K	Average load balance Δ							
	2	4	8	16	32	64	128	256
Random	1.00	1.01	1.02	1.04	1.08	1.11	1.2	1.26
DG	1.22	1.58	2.25	3.41	6.10	11.6	21.0	43.8
LDG	1.00	1.00	1.00	1.00	1.00	1.00	1.00	1.01
CDG	1.00	1.00	1.00	1.00	1.00	1.00	1.00	1.01
Metis	1.03	1.03	1.02	1.02	1.02	1.02	1.03	1.02

Table 4 Detailed results on Slashdot0811

K		2	4	8	16	32	64	128	256
Random	γ	45.7	68.6	80	85.8	88.6	90	90.8	91.1
	γ_c	1.7	4.5	8.8	15.2	24.3	36.4	50.2	63.5
	Δ	1.0	1.0	1.0	1.0	1.1	1.1	1.1	1.2
DG	γ	10.1	15.8	25.1	42.6	37.2	35.5	47.5	34.2
	γ_c	1.2	2.4	3.9	6.8	7.0	7.7	10.4	8.4
	Δ	1.6	2.6	4.4	6.0	12.8	28.8	39.8	117
LDG	γ	35.3	54.5	64.1	70.3	73.4	75.8	77	77.9
	γ_c	1.2	3.4	6.7	12	19.5	29	39.7	50.2
	Δ	1.0	1.0	1.0	1.0	1.0	1.0	1.0	1.0
CDG	γ	37	53.8	63.3	72.9	79.3	82.2	84.1	85.4
	γ_c	1	2.9	5.9	9.9	15.3	21.9	29.2	37
	Δ	1.0	1.0	1.0	1.0	1.0	1.0	1.0	1.0
Metis	γ	22.6	40.1	51.5	58.2	62.9	65.6	68	70.2
	γ_c	0.9	2.5	5.2	9.6	15.9	24.3	34	43.4
	Δ	1.0	1.0	1.0	1.0	1.0	1.0	1.0	1.0

6.3 Partition Results Using ADG Heuristic

In this section, we present the partition results using *ADG* on the same dataset (Table 1) and compare it with Random, *LDG* and *Metis*. We did not compare with *DG* since it is an unbalanced method. We take random method as the reference and use relative cost ratio ρ to evaluate other algorithms utilizing the affine message combining cost function (Eq. 5).

Specifically, to validate the performances on different settings of w_1, w_2 and w_3 described in definition (4), we set $w_3 = 1$ and use another symbol $\eta = w_1/w_2 = w_2/w_3$ to adjust the relative ratio between w_1, w_2 and w_3. η is greater than 1 since $w_1 \geq w_2 \geq w_3$ as we discussed in Sect. 4.1. Here we define η is ranged from 1 to 128 with two-fold increase in each step.

Experiments are performed on fixed number of partitions ($k = 256$).

As shown in Table 5, *ADG* outperforms the online method *LDG* on all the setting of η. The gap between *ADG* and *LDG* is increasing as the η. When η is larger

Table 5 Average relative cost ratio ρ of methods on 256 partitions using different combinations of w_1, w_2 and w_3

η	Relative cost ratio ρ							
	1	2	4	8	16	32	64	128
Random	1	1	1	1	1	1	1	1
LDG	0.72	0.75	0.80	0.83	0.86	0.88	0.89	0.90
ADG	0.71	0.69	0.64*	0.60*	0.57*	0.56*	0.55*	0.54*
Metis	0.57*	0.61*	0.65	0.70	0.72	0.74	0.75	0.76

"*"indicate the best average relative cost ratio

than 16 (their gap is 29 % then), their gap is still increasing but the increasing speed is reduced. At the same time *ADG* even outperforms *Metis* when η is larger than 4 and it reduces 20 % cost than *Metis* when η is 64.

6.4 Runtime Efficiency

All algorithms have been implemented in C++, and all experiments were performed on Linux computing environment. Each experiment is performed on a single core with Intel Xeon CPU at 2.2 GHz, and 64 GB of main memory. Wall clock time includes only the algorithm execution time, excluding the required time to load the graph into memory.

According to Table 6, in terms of the time efficiency, CDG and ADG are comparable to other two online algorithms. Random method is the fastest method and Metis keeps near stable time consuming on different numbers of partitions.

6.5 Performance Comparison in Different Stream Orders

In a sense, a heuristic's performance is also influenced by the stream order. Except random stream order which is generally used as default setting, the other two typical stream orders, DFS (Depth-First Search) and BFS (Breath First Search), are also

Table 6 Runtime result of CDG and ADG on Amazon0601 with different numbers of partitions

K	Run time (second)							
	2	4	8	16	32	64	128	256
Random	1	1	1	1	1	1	1	1
DG	3	3	3	4	5	7	12	17
LDG	3	3	3	4	5	7	12	18
CDG	3	3	4	5	6	7	11	19
ADG	3	4	4	6	8	8	12	20
Metis	32	33	33	33	34	34	34	35

often used to evaluate a heuristic [18]. To probe such influence on our proposed heuristics, we compared the performance of CDG and ADG and the competition heuristics on those three different stream orders: Random, DFS and BFS.

- Random: Vertices arrive according to a random permutation.
- DFS: We select a vertex uniformly at random. We perform depth-first search and sample the nodes with DFS. If there are multiple connected components, the component ordering is done at random.
- BFS: This ordering is identical to the DFS ordering, except that we perform breath first search.

(1) Comparison on average combining edge cut fraction (γ_c) and load balance (Δ)

Table 7 shows the average combining edge cut fraction (γ_c) and load balance (Δ) of five competition methods on the data in Table 1 and on eight different numbers of partitions (from 2 to 256).

As shown in Table 7, CDG outperforms the random and LDG methods on three stream orders. It has similar performance on random and DFS stream order and even better performance on BFS compared with Metis. DG method shows the worse performance on DFS and BFS stream orders.

(2) Comparison on average relative cost ratio ρ

To validate the performance of those heuristics in different stream orders according to the relative cost ratio ρ, we conduct the experiment on all the data described in Table 1. The number of partitions is fixed to 256.

Table 8 presents the average relative cost ratio ρ of four competition methods (DG method is not considered for it is an unbalanced heuristic) on fixed η (η = $w_1/w_2 = w_2/w_3 = 64$). ADG outperforms LDG and Metis on all the three stream

Table 7 Average γ_c and Δ on different numbers of partitions

Heuristic	Random		DFS		BFS	
	γ_c	Δ	γ_c	Δ	γ_c	Δ
Random	48.9		48.9		48.9	
DG	26.5		0		0	
LDG	32.9		30.0		28.7	
CDG	28.1		26.8		24.9	
Metis	25.6		25.6		25.6	

Table 8 Average relative cost ratio ρ of methods on 256 partitions using η = 64

Heuristic	Random	DFS	BFS
Random	1	1	1
LDG	0.79	0.66	0.69
ADG	0.64	0.55	0.57
Metis	0.67	0.67	0.67

orders. At the same time the performance of LDG and ADG in DFS and BFS orders is significantly better than that in random order. It can be ascribed to that in DFS and BFS orders that the heuristics of LDG and ADG are easier to allocate the nodes and their neighbours to the same partition and thus reduce the consumption on message channel.

Figure 4a–c presents the detailed results of the average relative cost ratio (ρ) on eight different η (1, 2, 4, 8, 16, 32, 64 and 128) in three stream orders. We can see that the average relative cost ratio of ADG is below than that of LDG in all stream orders. Moreover, ADG also shows better performance in all stream orders than Metis when η is larger than 4. Furthermore, the average relative cost ratio (ρ) of LDG and Metis is increased with η whereas that of ADG is stable. This further validates the advantage of adopting cost function with affine message combining in graph partition.

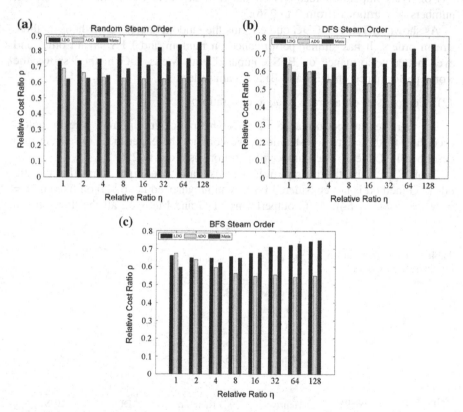

Fig. 4 Relative Cost Ratio (ρ) of Random, LDG and ADG in different stream orders. **a** Graph partitioning in random stream order. **b** Graph partitioning in DFS stream order. **c** Graph partitioning in BFS stream order

7 Discussion

In this paper, we propose a cost function for graph partitioning that considers not only message combining but also the different types of messages. Based on this cost function, we develop two heuristics CDG and ADG for large graph partitioning problem. We evaluate our method extensively on real-world graphs over a wide range of number of partitions and verify consistently the superiority of our method compared to existing ones. Specifically, despite the fact that CDG and ADG perform a single pass over the graph, it achieves performance comparable to and even better than Metis. Furthermore, our affine cost function can be dynamically adjusted to match the different realistic clusters to enhance the partition performance of the heuristics on different kinds of clusters.

Currently, CDG and ADG are the first heuristics that utilized different kinds of message combining cost functions in graph partition. They have the potential to further improve the performance by more elaborately analysing the relation between graph and the different types of combining messages.

Another interesting and potential future research problem is to derive an effective offline algorithm based on the affine cost function defined in this paper since it may be more fit for different messages consumed in modern clusters.

Acknowledgments The work described in this paper was supported by the US NSF grant CNS 1337899: MRI: Acquisition of Computing Equipment for Supporting Data-intensive Bioinformatics Research at the University of Kansas, 2013–2016.

References

1. Malewicz G, Austern MH, Bik AJ, Dehnert JC, Horn I, Leiser N et al (2010) Pregel: a system for large-scale graph processing. In: Proceedings of the 2010 ACM SIGMOD International conference on management of data, 2010, pp 135–146
2. Avery C (2011) Giraph: Large-scale graph processing infrastructure on Hadoop. In: Proceedings of Hadoop Summit. Santa Clara, USA
3. Zaharia M, Chowdhury M, Franklin MJ, Shenker S, Stoica I (2010) Spark: cluster computing with working sets. In: Proceedings of the 2nd USENIX conference on Hot topics in cloud computing,pp. 10–10
4. Low Y, Gonzalez J, Kyrola A, Bickson D, Guestrin C, Hellerstein JM (2010) Graphlab: a new framework for parallel machine learning. arXiv:1006.4990
5. Shao B, Wang H, Li Y (2013) Trinity: a distributed graph engine on a memory cloud. In: Proceedings of the 2013 international conference on Management of data, 2013, pp 505–516
6. Ke Q, Prabhakaran V, Xie Y, Yu Y, Wu J, Yang J (2011) Optimizing data partitioning for data-parallel computing. HotOS XIII
7. Kernighan BW, Lin S (1970) An efficient heuristic procedure for partitioning graphs. Bell Syst Tech J 49:291–307
8. Fiduccia CM, Mattheyses RM (1982) A linear-time heuristic for improving network partitions. In: 19th Conference on Design Automation, 1982, pp 175–181
9. Feige U, Krauthgamer R (2002) A polylogarithmic approximation of the minimum bisection. SIAM J Comput 31:1090–1118

10. Ng AY (2002) On spectral clustering: analysis and an algorithm
11. Karypis G, Kumar V (1998) A fast and high quality multilevel scheme for partitioning irregular graphs. SIAM J Sci Comput 20:359–392
12. LÜcking T, Monien B, Elsässer R (2001) New spectral bounds on k-partitioning of graphs. In: Proceedings of the thirteenth annual ACM symposium on Parallel algorithms and architectures, pp 255–262
13. Abou-Rjeili A, Karypis G (2006) Multilevel algorithms for partitioning power-law graphs. In: 20th International parallel and distributed processing symposium, 2006. IPDPS 2006, 10 pp
14. Dean J, Ghemawat S (2008) MapReduce: simplified data processing on large clusters. Commun ACM 51:107–113
15. Kang U, Tsourakakis CE, Faloutsos C (2009) Pegasus: a peta-scale graph mining system implementation and observations. In: Ninth IEEE International conference on data mining, 2009. ICDM'09, 2009, pp 229–238
16. Gonzalez JE, Low Y, Gu H, Bickson D, Guestrin C (2012) Powergraph: Distributed graph-parallel computation on natural graphs. In: Proceedings of the 10th USENIX symposium on operating systems design and implementation (OSDI), 2012, pp 17–30
17. Bourse F, Lelarge M, Vojnovic M (2014) Balanced graph edge partition
18. Stanton I, Kliot G (2012) Streaming graph partitioning for large distributed graphs. In: Proceedings of the 18th ACM SIGKDD international conference on Knowledge discovery and data mining, 2012, pp 1222–1230
19. Tsourakakis C, Gkantsidis C, Radunovic B, Vojnovic M (2014) Fennel: Streaming graph partitioning for massive scale graphs. In: Proceedings of the 7th ACM international conference on Web search and data mining, 2014, pp 333–342
20. Leskovec J, Krevl A (2014) SNAP Datasets: Stanford Large Network Dataset Collection

Big Data Analytics Platforms
for Real-Time Applications in IoT

Yogesh Simmhan and Srinath Perera

Abstract Big data platforms have predominantly focused on the volume aspects of large-scale data management. The growing pervasiveness of Internet of Things (IoT) applications, along with their associated ability to collect data from physical and virtual sensors continuously, highlights the importance of managing the velocity dimension of big data too. In this chapter, we motivate the analytics requirements of IoT applications using several practical use cases, characterize the trade-offs between processing latency and data volume capacity of contemporary big data platforms, and discuss the critical role that *Distributed Stream Processing* and *Complex Event Processing* systems play in addressing the analytics needs of IoT applications.

1 Introduction

Internet of Things (IoT) is an emerging architectural model that allows diverse sensors, controllers, devices, and appliances to be connected as part of the wider Internet [1]. IoT is driven by the growing prevalence of network-connected devices or "Things" that are present as part of physical infrastructure (e.g., *Smart Meters that connect to Power Grids and Smart Appliances at residences* [2]), which observe the natural environment (e.g., *Air Quality Monitors in Urban Settings* [3]), or monitor humans and society at large (e.g., *FitBit for individual fitness tracking and Surveillance Cameras at event venues*). In addition to monitoring and transmitting the observations over the network, these "Things" may also be controlled remotely, say to turn off a smart air conditioning unit or to turn on a smart vehicle, through signals over the network. As such, estimates on the number of such things

Y. Simmhan (✉)
Department of Computational and Data Sciences, Indian Institute of Science,
Bangalore, India
e-mail: simmhan@cds.iisc.ac.in

S. Perera
WSO2, Colombo, Sri Lanka
e-mail: srinath@wso2.com

© Springer India 2016
S. Pyne et al. (eds.), *Big Data Analytics*, DOI 10.1007/978-81-322-3628-3_7

115

will be part of the IoT range in the billions, and these things may be generic, such as a smartphone, or specialized to a domain, as in the case of a smart cart in a grocery store [4]. Some definitions of IoT include Humans as first-class entities within the internet of things and humans (IoTH) [5].

IoT enables the generation of enormous volumes of data observed from the sensing devices, and this begets a key question of how to meaningfully make use of this information. The promise of IoT lies in being able to optimize the network-connected system or improve the quality of life of humans who use or interact with the system. However, this is predicated on being able to analyze, understand, and act on this information. This opens up fundamental questions on big data analytics that are still in the process of being characterized [6]. Such analytics may be simple correlations between the outside air temperature and the load on the power grid (i.e., hot days lead to increased use of air conditioners, and hence greater load on the power grid), to more complex causal relationships, such as an evening accident on a freeway leading to a progressive neighborhood which delays the jump in power usage caused by electric vehicles being recharged when commuters reach home.

The ability to observe IoT systems does not mean that we can understand or reason about such complex systems, just as in the "butterfly effect," the ability to observe butterflies does not allow us to link the flapping of their wings to a future tornado at a remote location [7]. However, the unprecedented ability to sense physical and natural systems does offer a unique opportunity to apply, extend, and invent data analytics techniques, algorithms, and platforms to help make decisions to improve and optimize such systems. In particular, it places emphasis on the *Velocity* dimension of big data [8] and consequently, on the temporal aspect of data generation and its analysis. This is intuitive. IoT devices are often active continuously, leading to a perpetual stream of time series data being emitted from them. Such temporal data are transient in value and IoT systems benefit if this data is analyzed and acted upon in real time to close the loop from network to knowledge.

In the rest of the chapter, we introduce use cases and scenarios of emerging IoT domains that are grappling with big data and have a tangible societal impact. We use these to motivate specific analytics and techniques that are required by such domains to operate effectively. Subsequently, we discuss big data platforms that are making headway in offering such tools and technologies, and provide case studies of such analytics in action for IoT applications. We finally conclude with open problems we foresee for big data analytics in IoT, as this nascent field matures beyond its hype.

2   IoT Domains Generating Big Data

2.1   Smart Cities

There is intense global interest in enhancing the quality and sustainability of urban life. This is driven by the growing urbanization in developing countries, particularly highly populated ones like China, India, and Brazil, which is stressing the urban infrastructure and affecting the livability of residents, as well as the need to seamlessly integrate the urban ecosystem with a technology-driven lifestyle. The *Smart Cities* concept attempts to infuse sensing, control, and decision-making into different utilities offered in cities, such as smart transportation, smart power grids, and smart water management [3]. In this day and age, it also attempts to improve safety and health through urban surveillance and environmental monitoring for air and noise pollution.

Smart power grids monitor the consumers of power using advanced metering infrastructure (AMI), also called *Smart Meters*, to provide real-time information about the amount of power consumed, typically at 15-min intervals [9]. This, combined with metering at neighborhood transformers and community sub-stations, allows the power utility to get a *realistic view* of their distribution network. The goal for the utility is to use this data to manage their supply and demand to avoid a mismatch that could cause *brownouts and blackouts*, as well as to switch their supply mix from reliable but polluting coal and gas-based power plants to less-reliable solar- and wind-based generation. Data-driven targeting of specific consumers helps in reducing consumption through *demand-response optimization*, rather than increase production, which is another goal. This requires intelligent shifting, shaving, and shedding of loads from household appliances, electric vehicles, and industrial units based on their load profiles. Smart water management likewise attempts this intelligent resource management for water, with the additional goal of ensuring adequate water quality to mitigate health concerns.

Smart transportation [10] uses sensors that monitor road traffic conditions, using inductive-loop traffic detectors on roads, traffic cameras, and in-vehicle and in-person monitoring devices, with the goal of using such data to efficiently managing traffic flow through smart traffic signaling. This also extends to managing public transport by optimizing the schedule of buses and trains on-demand, based on real-time data, and planning the routing of freight vehicles in transport hubs and port cities to ease the traffic. Such data collection, particularly through surveillance cameras, also helps with vehicular safety (tracking hit-and-run accidents, or automobile theft), public safety (burglaries or attacks), and disaster management (tornadoes, earthquakes, urban flooding). Users already see the benefits of smart transportation through apps like Waze and Google Maps that offer directions based on real-time traffic conditions monitored through their smartphone apps, but getting a more integrated view over diverse sensor and human data for smart transport management is a particular challenge.

2.2 Smart Agriculture

With global warming affecting food supply and the increasing affordability of the world's population to food affecting its demand, there is starting to be a stress on the agricultural output. While mechanized farming by industry-scale farmers is the norm in developed countries, a majority of farming in developing countries is still human-intensive with low to negative profit margins. As such, technology and particularly IoT can play a role in improving practices at both these scales and maturity levels [11].

One of the key challenges in farming is to decide when to water the crops. Cash crops like vineyards are very sensitive to the soil moisture and humidity. The quality of the produce is affected by over or under irrigation, which also depends on weather conditions such as sunlight and warmth, rainfall, and dew conditions [12]. On the other hand, irrigation in developing countries like India relies on pumping groundwater, and the frequency of this depends on availability of intermittent electricity supply to operate pumps. As such, IoT can make it possible to use soil moisture data from ground sensors, remote sensing data from satellite imagery, and data from weather prediction models as well as supply schedule of power utilities to intelligently plan the irrigation of crops.

2.3   Smart Health and Lifestyle

Sports provide natural use cases for IoT, which, while less critical than other social scenarios, do provide early insight into emerging ideas due to rapid technology penetration. Common uses of IoT in sports rely on sensors placed in a player's shoes, helmet, or clothing, which provide high-resolution data (e.g., x, y, z location, speed, acceleration) about the player's actions coupled with vitals like heart rate. For example, the DEBS 2013 Conference's Grand Challenge [13] is based on a data collected from players' shoes and the ball in a soccer game. Also, American football teams have started placing cameras and sensors in players' helmets to detect concussions, and one can even buy basketballs off-the-shelf with sensors embedded within to track plays. The potential benefits of having such fine-grained data on the players and the equipment can ensure player's safety from injuries suffered at game time, better referee decisions, data-driven player selection (e.g., MoneyBall [14]), augmented TV broadcast with the enhanced game analysis, and even embedded virtual reality views for the audience.

Health and lifestyle examples range from activity monitors and smart watches such as *FitBit*, *Withings*, and *Apple Watch* to in-home care for the elderly. There is a growth in electronic devices that track people's behavior and basic health metrics, and enables them to get warnings about potential illness or a health condition, or just for personal analytics that help people reach fitness goals as part of a "quantified self" movement [15]. The increasing prevalence of smart watches and smart health devices is leading to individual health data being collected, and generic or specialized apps that can monitor and analyze them.

At the same time, such inexpensive sensors would make it possible to have sustainable in-home care for the elderly, recuperating patients, and those with long-term medical conditions to live at their home while having their health-monitored remotely [16]. Also IoT devices can improve the quality of hospital care and closer integration of case handling by monitoring medication to ensure patients are given the dosages consistent with their prescription, and avoid nosocomial ailments by ensuring caregivers wash their hands after procedures. In developing countries that rely largely on tertiary care by community nurses, neo-natal monitoring through bracelets coupled with smartphones can help detect if a baby shows signs of trauma that needs immediate medical attention [17].

2.4 Smart Retail and Logistics

Retail and logistics domains have a vital need to track their supply chain activity. For example, retailers are interested to track their inventory, shipping, and even the behavior of customers in their stores. RFID tags have been a major part of the supply chain for a while, with Walmart using them to handle their logistics [18]. It enables them to automatically track what items move in and out of the store without having to scan them, and to reliably know where each item is, avoiding operator errors. More generally, "Smart things" like RFID tags, GPS trackers, *iBeacons*, etc., can track items that are being transported and avoid costly errors in domains like airline baggage, postal vehicles, and even livestock and wildlife.

In logistics, speed and accuracy are vital. Within a retail store, smart tracking can reveal a wealth of information about consumer behavior and provide a rich interactive experience. Strategically placed sensors in store shelves and aisles can track what area of the store gets most attention and what regions confuse shoppers [19]. One often-overlooked sensor that is present in most stores is a video camera, typically used for security but which can, through streaming video analysis, also reveal customer behavior. On the other hand, mobile devices and Bluetooth low-energy (BLE) beacons can provide an interactive experience to the consumer, greeting them by name, helping them locate items, and even making suggestions on related products and discount coupons. These can provide mutually beneficial value to consumers and retailers, increase the sales volume for retailers, and help with efficient inventory management.

Despite their huge potential, IoT devices and the communication technology that connect them are just enabling tools—only as effective as how they are used and the decisions that they empower. As a result, deriving actionable insights from data collected through IoT devices and carrying out appropriate actions though IoT devices or by other means is an open challenge for IoT-based systems.

3 Role of Big Data Analytics in IoT

The above scenarios present an overview of the potential application domains that would benefit from big data being generated through IoT infrastructure. Next, we explore in greater detail the role of analytics and decision systems in these use cases.

IoT systems are an example of *autonomic systems* that garnered significant attention in the early 2000 s [20]. Autonomic system design has resulted in several types of control loops, which often follow the MEAP model of *Monitor, Analyze, Plan*, and *Execute* operations. This is similar to the *Observe, Orient, Decide* and *Act* (OODA) control loop used in other domains. These are often closed loops, since the *Execute* or *Act* step would likely cause the environment to change, which would trigger the loop again, and so on.

These feedback and control models are relevant within IoT applications as well. For example, a smart power grid application that performs home energy management may *monitor* the power consumption in the household and *observe* that the residents have just entered their home (e.g., the garage door opens), *analyze* their preferences, and *decide* to start the air conditioning unit that is configured to the optimum temperature level and even *plan* for coffee to be brewed in 15 min, by "talking" to smart appliances. Some of these scenarios are indeed possible, even at the present.

However, there are many challenges in building intelligent systems that can make automated decisions and executing them. Some of the questions to consider are:

1. How does it effectively *learn* from data, and dissociate signal from noise?
2. How can it *integrate* expert knowledge with observed patterns?
3. How can it understand the *context* (Where, When, Who, Where) and act accordingly?
4. How can it comprehend the *consequences* of and interference between different actions?
5. How does it plan for *causality that are not instantaneous*, but take place over time, across control iterations, and can fail?

Of course, these are challenging problems not unique just for IoT but fundamentally to many domains that can benefit from artificial intelligence, machine learning, and expert systems. However, the availability of big data in IoT domains makes them a particularly interesting candidate for exploration.

There is a great deal of diversity in IoT use cases and their underlying decision systems, and these systems frequently contend with the above challenges. We can classify *decision systems* based on increasing levels of complexity. While this classification is not meant to be comprehensive, it illustrates the different degrees of control capabilities that are possible, and analytics that are required.

1. **Visual Analytics**: Such techniques help humans analyze the data and present information to them in a meaningful form through a dashboard interface. They are designed to augment the human decision process with more information, presented in a cohesive and easily interpretable manner. An example is FitBit, which collects personal activity data using an in-person device and presents a summary of a user's effort levels, estimated calories burnt, and progress toward monthly goals, through a mobile or web application.
2. **Alerts and Warnings**: These engines allow decision logic to be provided by end users, and then uses these to interpret and classify data that arrives in order to

raise alerts or warning. These perform a certain degree of automated predefined analysis to help *highlight* situations of interest, which becomes critical when users have to deal with large volumes of data. For example, an environmental monitoring system may track the level of pollution or chemicals in an industrial city and send notifications on health hazards or chemical leaks to the citizens.

3. **Reactive Systems**: Systems may also go one step further by taking concrete actions (beyond notifications) based on their decisions. Generally, they are designed with a rule-based language that describes actions to be carried out when certain conditions are met. For example, a smart lighting system may turn off the lights when nobody is present in a room. Here, there is a tight coupling between just two physical components—an infrared sensor and light bulbs, and along a single data dimension—the IR level that indicates human presence, i.e., the system is not aware of the bigger picture but just the local, unidimensional situation.

4. **Control and Optimize**: Control systems operate in a closed loop where decisions lead to actions that are instantaneous, with the possibility that the actions can fail to meet the optimization goal. Control loop decision systems attempt to optimize the behavior of specific variables and also consider failure cases when deciding action outcomes. As discussed before, such MEAP or OODA systems are capable of generating a plan of action, executing the action, observing the response in a control loop, and recovering from a failure to meet a goal. For example, many electro-mechanical systems such as cruise control in a car or even a simple air conditioner thermostat operate on such a closed loop to either maintain the speed of the car or the ambient temperature.

5. **Complex Systems**: Such systems understand the context and interaction between several decision loops and are capable of making high-level decisions that span multiple dimensions within a single domain. For example, systems that manage city traffic can discern interaction between multimodal transports such as Bus, Metro, and Trains, in scheduling the road and rail traffic signaling for efficient transit.

6. **Knowledge-driven Intelligent Systems**: Complex infrastructure has cross-domain impact, with decisions in one domain impacting the other. Knowledge-based intelligent systems attempt to capture the relationship between different domains, such as transportation and power, or environmental conditions and healthcare, and optimize the decisions across these domains. The knowledge base itself is often specified by experts, and these may partially guide automated "deep learning" of correlations and causations within and across domains.

7. **Behavioral and Probabilistic Systems**: Human being are an intrinsic part of IoT though they are often overlooked. As such, they are both sources of data and means for control, through messaging, suggestions, and incentives. Behavioral systems attempt to include human models as a part of the overall IoT system, with likely nonuniform behavior. As a generalization, probabilistic and fuzzy systems incorporate nondeterminism as an integral part of decision-making, which goes above and beyond failures (Fig. 1).

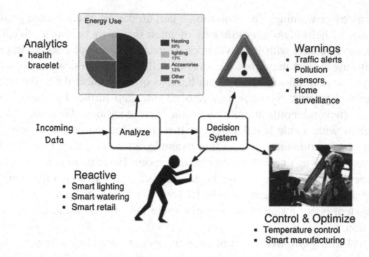

Fig. 1 Real-time decision-making

These systems vary from simple analytics performed in a batch mode, to real-time detection of patterns over data to issue alerts, to make complex plans and decisions that may have downstream consequences over time. The temporal dimension and the context are two particular characteristics that make these systems challenging. In the next sections we discuss some of these concepts in detail. However, behavioral systems and beyond are still in their infancy.

4 Real-Time Big Data Analytics Platforms

Decision systems that go beyond visual analytics have an intrinsic need to analyze data and respond to situations. Depending on the sophistication, such systems may have to act rapidly on incoming information, grapple with heterogeneous knowledge bases, work across multiple domains, and often in a distributed manner. Big data platforms offer programming and software infrastructure to help perform analytics to support the performance and scalability needs of such decision support systems for IoT domains.

There has been significant focus on big data analytics platforms on the volume dimension of big data. In such platforms, such as MapReduce, data is staged and aggregated over time, and analytics are performed in a batch mode on these large data corpus. These platforms weakly scale with the size of the input data, as more distributed compute resources are made available. However, as we have motivated before, IoT applications place an emphasis on online analytics, where data that arrives rapidly needs to be processed and analyzed with low latency to drive autonomic decision-making.

4.1 Classification of Platforms by Latency and Throughput

In Fig. 2, we classify the existing big data platforms along two dimensions: the average *throughput* of data processed per unit time (along the *Y* axis) and the *latency time* to process a unit of data and emit a result (*X* axis). *Volume-driven platforms* such as MapReduce scale to terabytes of data, and on an average, are able to process 100 MB/s of data per second (top right of figure). But this is in a batch processing mode, so the minimum time taken to generate a result (even for small data sizes, such as in the bottom right of figure) is on the order of minutes. This is forced by the need to store data on distributed disks, which introduces I/O and network overheads but also ensures persistence of data for repeated analysis. As a result, the time between data arriving and useful decision being made is in the order of minutes or more.

Databases, both relational and NoSQL, also use disk storage but can be used with indexes to support interactive searches. This guarantees persistence while avoiding the brute-force table scans of MapReduce platforms. OLAP technologies (e.g., Pentaho) and projects like Apache Drill build a layer on top of databases for interactive processing. For large datasets, indexing needs to be done a priori, introducing some additional latencies. In such platforms, the scalability of data throughput is sacrificed (10 KB–1 MB/s) in return for faster query response times on the order of seconds to minutes.

Stream Processing systems are also called "Real-Time processing" although they are in fact only low-latency (but not real-time) processing, in the strict sense of the word. Such systems can process data as they arrive at high rates by only keeping a small subset of the data in memory or on disk for the duration of processing, and can produce results in milliseconds. Here, the incoming data is transient and once processed is discarded (or separately stored in a NoSQL database for offline processing). Such platforms may operate on a single machine or on distributed systems. These are the focus of further discussion in this chapter.

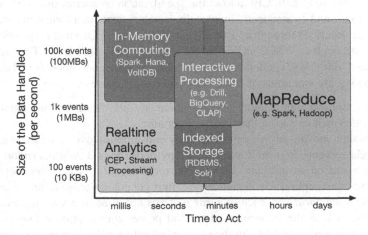

Fig. 2 Categorizing big data platforms based on latency of processing and size of data processed

Finally, *In-Memory Computing* uses the idea of loading all the data into distributed memory and processing them while avoiding random access to distributed disks during the computation. Platforms like Apache Spark offer a streaming batch model that groups events within a window for batch processing. This, while sacrificing some latency, helps achieve a higher throughput across distributed memory and can be 100's to 1000's of times faster than stream processing.

Among these technologies, streaming analytics plays a key role as a layer that can perform initial, low-latency processing of data available from IoT devices. Their results may directly lead to actions, stored for future processing, or feed into models that take decisions.

We next introduce specific technologies to support real-time analytics, and consider two use case scenarios as running examples. (1) *Sports Analytics for Soccer Games*, where sensors placed in the football players' boots and the goalkeeper's gloves generate event streams at 60 Hz frequency that contain time, location, and speed information. This is motivated by the DEBS 2013 Challenge [13]. (2) *Analytics for Smart Energy Management* in buildings that identify real-time opportunities for energy reduction using events from smart power meters and electrical equipment. This is motivated by the Los Angeles Smart Grid Project [9]. In the subsequent section, we discuss how these big data stream processing technologies are integrated into these use cases.

4.2   Online Analytics to Detect Exact Temporal Patterns

In many domains, the knowledge from domain experts and from past experiences provide us with interesting patterns that can occur, and such patterns (or signatures) may indicate situations of interest that require a response as part of a decision support system. We can detect the occurrence of such patterns and conditions using *Complex Event Processing* (*CEP*). CEP allows the specification of queries over one or more event streams, and consequently helps define such patterns in a formal manner. The queries are described through an Event Query Language (EQL) that is specific to the CEP engine, such as WSO2's CEP Siddhi language or Esper's EQL. The survey of Cugola and Margara [21] offers a detailed review of these query models and engines.

Following are examples of some such patterns of interest:

- *Filter* ambient temperature measurement events from an air conditioner (AC) unit if it crosses a threshold temperature.
- Maintain a *moving average* of the energy consumption over 1 h time windows.
- Correlate events from multiple streams by performing a *Join operation* across them based on an attribute, such as the room location of a power meter generating consumption events and an AC unit generating temperature events.
- Detect a *temporal sequence of event patterns* specified as a state machine over events, such as the occurrence of several power consumption events of a low value followed by a spike in their value indicating a heavier energy load.

- *Preprocess events*, such as transforming their units or dropping unused attributes of the events.
- Track the *change in state over space and time* of an entity, such as the location of a football.
- *Detect trends* captured by sequences, missing events, thresholds, outliers, and even complex ones like "triple bottom" used in algorithmic financial trading.

As a generalization of these examples, CEP systems support the following categories of *query operators*, shown with a sample CEP query definition using the Siddhi EQL.

1. **Filters and Transformations**—This filters events from an input streams based on a predicate and places them, possibly after some transformation, into an output stream.

   - ```
     from    PowerStream[consumptionKWH>10]    insert    into
     HighPowerStream
     ```
   - ```
     from    ACUnitStream#transform.Op((temperatureF-32)
     *5/9)
     insert into ACUnitMetricStream
     ```

2. **Window and Aggregation operations**—This operator collects data over a window (e.g., period of time or some number of events) and runs an aggregate function over the window.

   - ```
     from    PowerStream#window    (60    min)    select    avg
     (consumptionKWH)
     ```

3. **Join two event streams on an attribute**—This operator matches events from multiple data streams or data from the same stream based on a shared attribute value, e.g., location. This is similar to a database join, but performed on a window of events, e.g., of 1 min duration.

   - ```
     from PowerStream#window (1 min) join ACUnitStream
     on PowerStream.location = ACUnitStream.location
     ```

4. **Temporal sequence of event patterns**—Given a sequence of events, this temporal operator can detect a subsequence of events arranged in time where some property or condition is satisfied between the events, e.g., the consumption KWH between the events is greater than 10. This can also use regular expression matches between events, say, to detect a pattern of characters from a given sequence of characters.

   - ```
     from every e = PowerStream
     -> PowerStream[e.consumptionKWH - consumptionKWH > 10]
     insert into PowerConsumptionSpikeStream
     ```

Just as SQL provides a high-level abstraction over database queries, advanced CEP queries can be composed by combining the above four basic query types to
     ```

detect complex conditions. They also support the inclusion of user-defined functions as part of their transformations to allow more sophisticated and domain-specific operators like machine learning or forecasting algorithms to be embedded within the CEP processing. For example, the bounds generated by a regression tree learning algorithm can be used as a filter condition within a CEP query.

A *CEP Engine* is a middleware server that executes CEP queries over event streams and produces a resulting stream of output events. The engine constructs an execution graph based on a given user query, where nodes in the graph represent the basic operators (e.g., Filter, Window) and edges in the graph represent event streams. The output event stream of a basic query can form the input stream to another basic query, thus allowing complex CEP queries to be composed. When a new event arrives at a stream, the event is submitted to all operators attached to that stream. Each operator processes the event and places the result on its output streams, which will then be delivered to downstream operators to which those output streams are connected to.

CEP engines offer fast execution of CEP queries due to their in-memory model of execution that does not rely on disk storage. Each of basic operators is specifically optimized using specialized algorithms. Also, CEP engines rely on compact memory efficient and fast in-memory data structures. The fast execution allows for low-latency ("real-time") processing of input event stream so that the engine can keep up query execution even with input stream rates. The small memory footprint also ensures that window queries can span a wide time or event window. CEP engine also provides various other optimization, as discussed in literature [22], such as

1. Fast event flows between distributed operators,
2. Multi-threaded executions with minimal concurrency control required from the user,
3. Enforcement of event ordering even with data-parallel parallel execution,
4. Throttling and rate limiting, and
5. Checkpoints and recovery for resilience.

As an exemplar of how a CEP engine works, we briefly consider the Siddhi CEP engine. Siddhi queries are defined in the format illustrated earlier. The engine analyzes those queries to build a processing graph as discussed above. Input events are placed on incoming edges in the graph, the operator processes them and sends them to the next operator(s) if the condition matches, and eventually the final events arrive at the output edges of the graph. Operators are stateful. For example, a window operator will remember events it has received until conditions for the window trigger has been fulfilled. Most of the complexity of the CEP engine is in its operators, and they are crafted to be both CPU and memory efficient.

4.3 Distributed Stream Processing

While CEP allows analytics of interest to be captured and detected in real time from streams of events, the process of ensuring the event streams is made available in an appropriate form falls within the space of stream processing. Stream processing allows composition of real-time applications as a directed acyclic graph (DAG), where vertices are application logic tasks, while edges are streams of events.

Two distinctions between stream and complex event processing are that the events in CEP are strongly typed, and the queries conform to well-defined logic that operate on typed events and are visible to the CEP engine to perform optimizations. In stream processing, on the other hand, the event contents as well as the application logic are opaque to the stream processing engine, and the engine is primarily responsible for moving event payload between application logic and the execution of the logic over the events. The events are just a binary or text payload that the engine does not examine. In that sense, one can consider CEP as specialization of stream processing systems.

Distributed Stream Processing (*DSP*) has gained interest recently due to the increasing quanta of real-time data that arrives and the hardware limits of single-machine, shared-memory execution. In DSP, the tasks of the DAG can be distributed across different machines in a local area network (LAN) or a Cloud data center, with event streams passing between them. Apache Storm is an example of such a system.

DSP are faster than CEP and offer a lower latency for processing since they avoid the overheads of introspecting the contents of the message. Consequently, they are used as preprocessors for cleaning and ingesting data from sensors, and their outputs can be passed onto online analytics platforms such as CEP, or stored for offline analytics. Such preprocessing often takes the form of an extract–transform–load (ETL) pipeline used in data warehousing or extract–transform–analyze, except that it is done in near real time. These pipelines can help convert the format and representation of the data, detect and clean outliers, perform sanity checks and quality control, etc. on the incoming data. DSP also allow analytic logic to be embedded within them as a task, so that a CEP engine or a machine learning logic can be a task within a DSP application.

For example, in the figure below, data streaming from hundreds of building sensors in a binary format is parsed to *extract* relevant information, and then *transformed* into a standard format necessary for analysis, a copy *stored* in a NoSQL archive for future processing, while a duplicate copy is forked for real-time analytics by a *CEP engine* (Fig. 3).

Despite the events and task logic of the DAG themselves being opaque, DSP engines support several different *semantics* when composing streaming applications. These include whether to perform a *duplicate* or *interleave/split* of outgoing events from a task, and using transparent keys with opaque values to perform a *map* operation from one task to a set of downstream tasks. Tasks may also perform either a *merge* or a *join* of multiple incoming messages, where the former interleaves

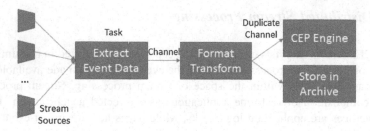

Fig. 3 Typical extract–transform–load–analyze model used for real-time data processing

incoming messages while the latter aligns messages based on arrival order. Tasks may be *stateful* or *stateless*, which determines whether the state from processing a message is available to upstream events that arrive. Often, DSPs *do not guarantee ordering* of events as distributed coordination is more difficult and favors faster execution instead.

The *selectivity* of a task determines the number of events it generates when consuming a single event, and its *latency* is an estimate of time taken to process one event. DSPs have a natural ability to *pipeline* execution of events. The *length* and the *width* of the DAG, respectively, determine the total latency of the *critical path* of execution, and the number of parallel tasks at the widest width of the DAG which can be exploited for *task parallelism*, if preceded by a duplicate task semantic. Stateless tasks can also leverage *data parallelism* by adding more processors to execute a single task, either on multiple threads in a machine, or across different machines in a cluster or virtual machines in the Cloud. This is more challenging for stateful tasks.

More recently, there is interest in distributing such stream processing engines to go beyond a single data center into distributed execution across edge devices and the Cloud. This *Edge + Cloud* model of execution is motivated by the growing prevalence of generating data from distributed IoT sensor devices that themselves have computing capabilities [23], albeit limited. Devices like smartphones, Raspberry Pi, and Arduino running mobile processors and hosting sensors can perform part of the data filtering on-board rather than push all data to the Cloud or to a centralized data center for processing. This has several benefits. It limits the need to transfer large quantities of data that can incur communication costs. It also reduces the network round-trip latency that is paid when sending the data to the Cloud for analysis and receiving a response to take action. Finally, it enhances the privacy of the data sources as it allows for local analysis on-site rather than pushing it to a public Cloud. At the same time, there may be some analytics such as aggregation and comparisons across sensors/devices that require limited event streams to be centrally accumulated for analysis. This distribution of the computation across edge and the Cloud to meet the different metrics of costs, time, and privacy is of growing research interest as IoT applications start to actively utilize DSPs.

5   Real-Time Analytics Case Studies

We next expand on the earlier example IoT domains of Smart Energy Management and Sports Analytics, and discuss how CEP and DSP systems are used for real-time analytics within these applications.

5.1   Analytics for Smart Energy Management

Smart power grids use information gathered from sensors that monitor the electricity transmission and distribution infrastructure to ensure reliable, efficient, and sustainable generation and supply of power. In addition to individual residences in a city-scale electric grid, there are also islands of *micro-grids* that are operated by large institutional campuses such as universities, office parks, and heavy industry. Such micro-grids control the power distribution within their campus, with local cogeneration as well as electricity supply from the city grid, and offer the ability to deploy IoT solutions to instrument, monitor, and control the micro-grid behavior in an intelligent manner.

The University of Southern California's (USC) Campus Micro-grid, part of the Los Angeles Smart Grid project, is an example of such a model where stream and CEP platforms are used [9] with similar efforts taking place at other institutional campuses worldwide. Diverse sensors to monitor the micro-grid range from *smart power meters* that measure KWh, power load, and quality, to *building area networks* that can monitor and control HVAC (heating, ventilation, air conditioning) units, lighting, and elevators. In addition, *organization, spatial and schedule information* on the departments, people, buildings, and class schedules are also available.

These provide a mix of slow (buildings, departments) and fast (sensor readings, schedules) changing data that has to be analyzed in real time for smart power management. Specifically, the analytics aims to perform *demand-response optimization*, wherein the power consumption of the micro-grid has to be curtailed *on-demand* when a mismatch between the available power supply and expected power demand is identified. Stream and complex event processing platforms play vital a role in this decision-making process.

(1) **Information integration pipeline**: Data coming from thousands of diverse sensors have to be preprocessed before analysis can be performed on them. Preprocessing includes extracting relevant fields from the sensor event stream, applying quality correction, performing unit transformations, and annotating them with static information available about that sensor. These steps are performed by an information processing pipeline that is composed using a distributed stream processing platform [24]. The outcome of the preprocessing is further forked into three paths: one, for semantic annotation and archival in a RDF data store for offline querying, analysis, and visualization; two, for

performing demand forecasting over KWh events that arrive from smart power meters; and three, for detecting power demand and curtailment situations using a CEP engine to aid decision-making.

(2) **Demand forecasting using time series**: One of the paths taken from the information processing pipeline is to predict future energy demand by each building in the micro-grid based on past behavior. For this, we use the ARIMA time series forecasting model that uses the recent history of KWh events from a building for its prediction. The ARIMA model is included as a logic block within the distributed stream processing pipeline and the forecasts are generated as a series of events. These are further analyzed in real time as part of the DSP to see if the sum of impending demand from different buildings is greater than the available generation capacity during that future period. This helps in deciding if a power curtailment action is required within the micro-grid.

(3) **Detecting power demand and curtailment opportunities**: Another event stream that forks from the preprocessing pipeline is to a CEP engine, which appears as a logic block within the DSP. This CEP engine has two types of analytics queries registered with it, one to *detect energy spikes* that may occur due to special circumstances that cannot be captured by the time series model, and another to *detect potential energy leaks* that can be plugged to reduce energy consumption when required. We offer a few examples of these two classes of CEP queries [25].

- *Find all labs where a class is scheduled to take place in the next 15 min and the current power consumption by that lab is* < 100 Wh. This indicates that the lab rooms can expect to draw more power once the class starts due to teaching equipment being turned on and a higher room occupancy. This queries the *classSchedule* event stream and the *smartMeter* event stream and can be implemented as a temporal sequence.

```
from c = classSchedule
   ->Time[c.time - now < 15m]
   ->smartMeter[c.bldNo == bldNo and power < 100kW]
insert into spikeStream
```

- *Find all buildings where the ambient temperature is below the setpoint temperature for the HVAC, and the ambient temperature has increased by more than 5'F over the past 1 h.* This indicates that the buildings where the HVAC is in low power mode may switch to a high power mode if the ambient temperature continues to rise. This queries the *roomTemperature* and the *HVACSetpoint* event streams and can be implemented as a temporal sequence.

```
from t = roomTemperature
   ->roomTemperature[at - at > 5]
   ->HVACSetpoint[t.at < setpointT]
insert into spikeStream
```

- *Find all labs where a scheduled class has completed 15 min ago, no classes are currently scheduled and the current power usage is* > 100 Wh. This indicates a potential for energy savings due to equipment that has not been turned off after a class has completed. This queries the *classSchedule* and *smartMeter* event streams. Here the NOT operator (!) checks if such an event has not happened.

```
from c = classSchedule
   ->t = Time[now - c.time > 15m]
   ->!classSchedule[startT <t.now and endT > t.now]
   ->smartMeter[c.bldNo == bldNo and power > 100kW]
insert into curtailmentStream
```

- *Find all buildings where more than 6 HVAC units are active at the same time for more than 10 min each.* This queries the *HVACStatus* event stream to identify buildings where HVAC units can be duty-cycled to operate in a round-robin fashion, thereby flattening the energy load across time. This is handled with two queries.

```
from u = HVACStatus[action=on]
   ->!HVACStatus[unitid == u.unitid and action=off]
   ->t = Time[time - u.time > 10m ]
select u.bid, u.uid, t.now
insert into HVACUnitOn10m

from HVACUnitOn10m
select bid, count(uid) as count group by bid
   having count > 6
insert into curtailmentStream
```

5.2 Sports Analytics

The Distributed Event-Based Systems (DEBS) Conference 2013's Event Processing Grand Challenge offers a novel IoT use case for soccer analytics that is addressed using CEP platforms. Data for this use case came from a Soccer game that was played with sensors placed in the ball, the players' boots, and the goal keeper's gloves. Each sensor generates events which describes the current location (x, y, z), the timestamp, velocity (vx, vy, vz), and acceleration (ax, ay, az), at the rate of 60 Hz.

This data can provide complete information of the game play. However, understanding and deriving higher levels events such as kicks, passes, ball possession, and offside requires detecting complicated temporal queries from these raw events. As part of the challenge, we implemented those queries using the Siddhi Complex Event Processing engine, and achieved throughput in excess of 140,000 events/sec [26]. We look at some example queries to understand how the aforementioned CEP operators are combined to detect such complex sport strategies.

Let us take *Ball Possession* as an example. In a soccer game, it is not fair to measure players just by the number of goals they have scored as defensive players get fewer chances to score. Ball possession, the time duration for which a player controlled the ball, is a much better predictor of a player's performance.

To detect ball possession, we first need to detect *kicks on the ball*. To do that, we used the hypothesis that if the player and the ball are within one meter of each other, and the ball's acceleration has increased by more than 55 m/s^2, then player has kicked the ball. This condition can be detected by combining the player position event stream and the ball position event stream using a *join query* as follows. Note that *getDistance()* is a user-defined function that uses a pair of (x,y,z) locations to determine the Cartesian distance between the locations, in meters.

```
from Ball#window.length(1) as b join
     Players#window.length(1)  as p unidirectional
     on debs:getDistance(b,p)< 1000 and b.a > 55
select ...
insert into kickStream
```

Once we have detected kick events that have occurred on the ball by specific players, we can detect the ball possession by each player. Player A has *ball possession* for the duration between the first time that s/he kicked the ball, followed by any number of kick events by the same Player A, and terminated by a kick event by some other player. This can be detected using a temporal event sequence on the *kickStream* as given below. This query is a temporal sequence where commas after 'from' separate an event or a condition about an event in the sequence.

```
from old = kickStream,
     b = kickStream[old.pid != pid ],
     n = kickStream[b.pid == pid]*,
     e1 = kickStream[b.pid != pid ] or
     e2 = ballLeavingKickStream
select ...
insert into ballPossessionStream
```

Such analysis can variously be used for sports management studies on the effectiveness of game play and the players, to build a dashboard on game status, to generate alarms about critical events in the game (e.g., missed offside), and real-time game analysis and predictions about the next move.

6 Discussion and Open Problems

The use cases we have presented exemplify the rich and novel information space that IoT applications offer for analytics and decision-making. These have a tangible impact on both technology and human lifestyle. Real-time big data platforms such as distributed stream processing and complex event processing that we have discussed are two of many platforms that can help coordinate, process, and analyze such information to offer actionable intelligence that offer the "smarts" to these IoT applications. However, as we have observed, these platforms need to work in tandem with other advances in computing, in areas like machine learning, data mining, knowledge harvesting, deep learning, behavioral modeling, etc. to provide holistic solutions that are robust and sustainable.

There are several open challenges on fast data processing platforms themselves. Incorporating semantics into processing events is important to bring in contextual information about diverse domains into the queries and combine offline knowledge bases with online data that stream in. Such contextual information include schedule and environmental information, habits, and interests that are learned, and proximity with other entities in the virtual and physical space that help integrate with an active world of humans, their agents, and things.

While the scalability of high-velocity big data platforms on captive commodity clusters is well studied, making use of elastic Cloud computing resources to on-demand adapt the execution to runtime changes in the application logic or incoming data rates is still being investigated. The nominal utility cost of stream and event processing and the real cost paid for Cloud computing resources will also come into play.

Smart devices that we carry on our person or are deployed on physical infrastructure have capable computing power despite their low power footprint. Enabling DSP and CEP engines to effectively leverage their compute capability, in conjunction with more capable computing resources at centralized data centers and Clouds, is important. This is sometimes referred to as fog computing or mobile Clouds. This has the advantage of reducing the round-trip latency for processing events that are often generated by the sensors, and actions taken based on the events have to be communicated back to actuators that are collocated with the sensors. This can also be robust to remote failures by localizing the closed-loop decision-making. Further, issues of data privacy can also be handled in a personalized manner by offering the ability to select the computing resource to perform the analytics and decision-making on.

Decisions that are driven by these analytics can have an immediate effect or may impact future actions. Defining performance requirements for such fast data processing platforms that take into account the lead time for decision-making and the resulting actions themselves can help prioritize processing of information streams. Having policy languages that can capture such dependencies, and real-time processing engines that can interpret and adapt to such needs, will help.

There are several open challenges specific to CEP engines too. First is scaling CEP execution beyond single machine. Several systems use data partitions, either derived automatically or given explicitly by the users, to break the execution into multiple nodes [22, 27]. For IoT use cases, handling out of order events is a major challenge. With thousands of sensors without a globally synchronized clock, events can arrive from sensors out of sync, and many CEP operators that are sensitive to the event ordering will find it challenging to enforce order and also ensure runtime scalability. Google's MillWheel is an effort to address this problem [28]. Finally, it is unavoidable that some of the sensors or readings may be faulty in an IoT deployment. Taking decisions despite those uncertainties is another challenge [29, 30].

References

1. Gubbi J, Buyya R, Marusic S, Palaniswami M (2013) Internet of Things (IoT): a vision, architectural elements, and future directions. Fut Gen Comput Syst 29(7):1645–1660. ISSN:0167-739X, http://dx.doi.org/10.1016/j.future.2013.01.010
2. Siano P (2014) Demand response and smart grids—A survey. Renew Sustain Energy Rev 30:461–478. ISSN:1364-0321, http://dx.doi.org/10.1016/j.rser.2013.10.022
3. Zanella A, Bui N, Castellani A, Vangelista L, Zorzi M (2014) Internet of things for smart cities. IEEE Internet Things J 1(1):22–32. doi:10.1109/JIOT.2014.2306328
4. Gartner Says 4.9 Billion connected "Things" will be in use in 2015. Press Release. http://www.gartner.com/newsroom/id/2905717. Accessed 11 Nov 2015
5. #IoTH: The internet of things and humans, Tim O'Reilly. O'Reilly Radar. http://radar.oreilly.com/2014/04/ioth-the-internet-of-things-and-humans.html. Accessed 16 April 2014
6. Barnaghi P, Sheth A, Henson C (2013) From Data to Actionable Knowledge: Big Data Challenges in the Web of Things [Guest Editors' Introduction]. IEEE Intell Syst 28(6):6, 11. doi:10.1109/MIS.2013.142
7. Lorenz E (1972) Does the flap of a butterfly's wings in Brazil set off a tornado in Texas? AAAS
8. Laney D (2001) 3D data management: controlling data volume, velocity and variety. Gartner
9. Simmhan Y, Aman S, Kumbhare A, Rongyang L, Stevens S, Qunzhi Z, Prasanna V (2013) Cloud-based software platform for big data analytics in smart grids. Comput Sci Eng 15 (4):38,47. doi:10.1109/MCSE.2013.39
10. El Faouzi N-E, Leung H, Kurian A (2011) Data fusion in intelligent transportation systems: progress and challenges—A survey. Inf Fusion 12(1):4–10. ISSN:1566-2535, http://dx.doi.org/10.1016/j.inffus.2010.06.001
11. Ma J, Zhou X, Li S, Li Z (2011) Connecting agriculture to the internet of things through sensor networks. Internet of Things (iThings/CPSCom). In: International conference on cyber, physical and social computing, pp 184,187, 19–22 Oct 2011. doi:10.1109/iThings/CPSCom.2011.32
12. Serrano L, González-Flor C, Gorchs G (2010) Assessing vineyard water status using the reflectance based Water Index. Agric Ecosyst Environ 139(4):490–499. ISSN:0167-8809, http://dx.doi.org/10.1016/j.agee.2010.09.007. Accessed 15 Dec 2010
13. The ACM DEBS 2013 Grand Challenge. http://www.orgs.ttu.edu/debs2013/index.php?goto=cfchallengedetails
14. Lewis M (2003) Moneyball: the art of winning an unfair game. W. W. Norton & Company
15. Adventures in self-surveillance, aka the quantified self, aka extreme Navel-Gazing, Kashmir Hill. Forbes Mag. Accessed 7 April 2011
16. Suresh V, Ezhilchelvan P, Watson P, Pham C, Jackson D, Olivier P (2011) Distributed event processing for activity recognition. In: ACM International conference on Distributed event-based system (DEBS)
17. Rao H, Saxena D, Kumar S, Sagar GV, Amrutur B, Mony P, Thankachan P, Shankar K, Rao S, Rekha Bhat S (2014) Low power remote neonatal temperature monitoring device. In: International conference on biomedical electronics and systems (BIODEVICES), 3–6 March 2014
18. Malone M (2012) Did Wal-Mart love RFID to death? ZDNet. http://www.zdnet.com/article/did-wal-mart-love-rfid-to-death/. Accessed 14 Feb 2012
19. The Nexus of Forces in Action—Use-Case 1: Retail Smart Store, The Open Platform 3.0™ Forum. The Open Group. March 2014
20. Kephart JO, Chess DM (2003) The vision of autonomic computing. IEEE Comput 36 (1):41–50. doi:10.1109/MC.2003.1160055
21. Cugola G, Margara A (2012) Processing flows of information: from data stream to complex event processing. ACM Comput Surv 44, 3:Article 15

22. Jayasekara S, Kannangara S, Dahanayakage T, Ranawaka I, Perera S, Nanayakkara V (2015) Wihidum: distributed complex event processing. J Parallel Distrib Comput 79–80:42–51. ISSN:0743-7315, http://dx.doi.org/10.1016/j.jpdc.2015.03.002

23. Govindarajan N, Simmhan Y, Jamadagni N, Misra P (2014) Event processing across edge and the cloud for internet of things applications. In: International Conference on Management of Data (COMAD)

24. Wickramaarachchi C, Simmhan Y (2013) Continuous dataflow update strategies for mission-critical applications. In: IEEE International Conference on eScience (eScience)

25. Zhou Q, Simmhan Y, Prasanna VK (2012) Incorporating semantic knowledge into dynamic data processing for smart power grids. In: International semantic web conference (ISWC)

26. Perera S (2013) Solving DEBS 2013 grand challenge with WSO2 CEP/Siddhi. Blog Post. http://srinathsview.blogspot.in/2013/05/solving-debs-2013-grand-challenge-with.html

27. Wu S, Kumar V, Wu KL, Ooi BC (2012) Parallelizing stateful operators in a distributed stream processing system: how, should you and how much? In: ACM International conference on distributed event-based systems (DEBS). http://doi.acm.org/10.1145/2335484.2335515

28. Akidau T et al (2013) MillWheel: fault-tolerant stream processing at internet scale. In: Proceedings of the VLDB endowment, pp 1033–1044

29. Skarlatidis A (2014) Event recognition under uncertainty and incomplete data. Doctoral thesis, University of Piraeus

30. Wasserkrug, S et al (2008) Complex event processing over uncertain data. In: International conference on Distributed event-based systems (DEBS). ACM

Complex Event Processing in Big Data Systems

Dinkar Sitaram and K.V. Subramaniam

1 Introduction

Complex event processing can be applied to solve problems arising in many interdisciplinary areas of computing where the data is obtained from different sources and at differing granularities both in time and space. This is different from a traditional store and process architecture where the data is first stored in a database and then analyzed. Given the extremely high data rate that some of the *Big Data* applications produce, the event processing model has provided an alternative to the store process model. Twitter, for example, has replaced the store process model by a distributed content-based *publish–subscribe* model [16]. Similarly, Google photon [5] mines event streams from user clicks and query terms to determine if an advertisement was associated with a search term; this is used to report ad clicks to the advertiser and helps Google monetize advertisement revenue.

The above examples are both from social media and search, and are examples of real-time stream processing with simple event models. In this chapter, we will look at an alternative big data application—manufacturing intelligence—as a motivational example to help illustrate the need for complex event processing. The tidal race data stream warehousing solution [22] considers the semiconductor manufacturing example and highlights the need for complex analytics processing on a real-time system. They argue on the need for weaker consistency in processing the real-time processing streams to make progress over time. In this chapter, we argue that there is great potential in integrating complex event processing models with stream processing to gain meaningful insights on the manufacturing process.

D. Sitaram (✉) · K.V. Subramaniam
PES University, Bangalore, India
e-mail: dinkar.sitaram@gmail.com

© Springer India 2016
S. Pyne et al. (eds.), *Big Data Analytics*, DOI 10.1007/978-81-322-3628-3_8

2 Manufacturing Intelligence

Consider the wafer cleaning process being utilized in the semiconductor industry [33]. This consists of passing the semiconductor wafer through a series of tanks as shown in Fig. 1. At each stage, the material is subject to cleaning at particular temperature and pressure with a chemical mixture. To ensure that the concentration of the chemicals is kept constant, the chemicals are replenished at periodic intervals possibly every 4–5 runs. Also, the temperature and pressure during a particular run are not necessarily kept constant and are varied based on the need of the process being employed. For example, it is possible that the temperature may be slowly raised to reach the required value over a period of a minute at a particular rate as illustrated in Fig. 2, but the pressure of liquid may be sprayed on the material in bursts at different rates across the time period. For example, the machine may be configured to spray two bursts in the first 10 s and then again fire three more bursts after a small gap as illustrated in Fig. 3.

The intensity and the timing of the bursts are both critical in ensuring that cleansing process is able to remove the impurities from the substrate. Further, given a large manufacturing system, there will be multiple different machines and each machine may have different parameters that the process control engineer would vary in order to obtain optimal yield. Any deviation from these could potentially lead to poor yield.

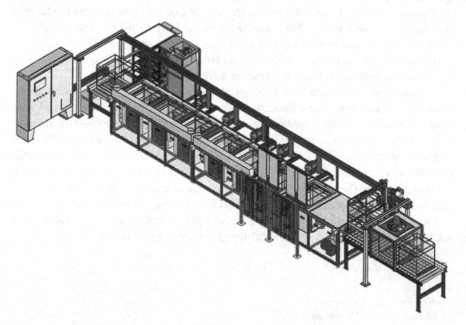

Fig. 1 Schematic of multi-tank cleaning system employed in the semiconductor industry

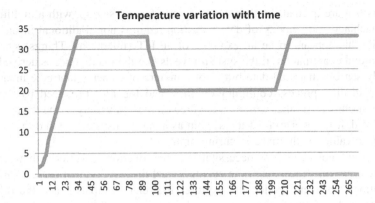

Fig. 2 Variation of temperature over time during a cleaning cycle (for illustration)

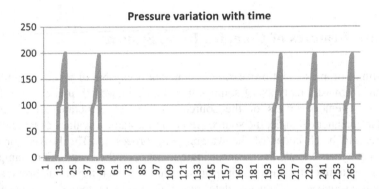

Fig. 3 Variation of pressure over time during a cleaning cycle (for illustration)

A manufacturing intelligence system must be able to receive data from various types of sensors (pressure, temperature and, liquid flow rate), identifier tags such as RFID sensors and bar code sensors that identify the substrate material and be able to notify the process engineers of any deviations from the process. It must additionally also be able to keep track of all materials that underwent the deviated process. Note that not only the variety of data is large, but also the granularity of relevant information is varied across time and space. For example, temperature/pressure sensors are valid only for one cleaning session. However, replenishment of the chemical mixture must happen across multiple cleaning cycles.

Traditional DBMS-based solutions will not be able to handle the volume and velocity of the data due to additional store/retrieve operation that is required to be performed in the critical path. For example, the temperature pressure sensors are typically required to be read every 50 ms, and a write operation across a network could potentially take longer than 50 ms thus building up a backlog of write requests that are queued. Further, given a large factory setting, with a multitude of such machines, the velocity of data ingestion requires non-traditional means.

While this appears to be an example of an IoT (Internet of Things) system, a more careful examination of the system reveals that the intelligence is derived from not only sensors that read data but also a mixture of cameras that take images for tracking quality, process configuration files, and log files produced by the controlling machines. In other words, given the volume and variety of the data that is to be ingested, it is better to treat the system as a big data system so that the analysis can derive value for the manufacturing unit.

The above considerations necessitate the introduction of systems that can process and draw inferences from the events in real time. As argued earlier, for many applications, an event model that captures the relationship between the different events is needed. In the next section, we give a brief overview of various features of such a complex event processing system.

3   Basic Features of Complex Event Systems

A real-time complex event processing system that is capable of processing millions of events from various types of sources that can be viewed as illustrated in Fig. 4 [12]. The event observers or the sources on the left generate events. In our multi-tank cleaner example, the sources refer to timestamped values of temperature, pressure, and batch codes of the wafers being processed. The brokers or event processing agents encode the business logic to act upon the events. For example, to decide if the temperature reached its critical value over an interval, the brokers have to process a sequence of values to determine if the operating temperature is valid. If so, it signals an event indicating the operating temperature which is valid. If not, an operating temperature invalid event is raised; this generated event can act as a source event for subsequent stages in the event processing network. Observe that the graph of the event processing network is logical and does not imply the physical distribution of the brokers. The processing agents can be distributed across a set of machines and need not be on a single machine. Also, this network must not have any cycles. There may be consumers that are interested in the events being generated by the system. These are represented by the sinks in Fig. 3.

A complex event processing system such as the one described above will have to consider the following aspects in designing the system. These are based on the eight requirements enumerated by [34].

1. **Data Model/Semantics**—this provides a mechanism to handle data from a variety of sources and also to handle relationships across various events. Consider our example of the manufacturing intelligence application. Herein, we need not only a model for being able to extract events from various types of sensors and log files, but also a model that can help express the relationship among the various events.

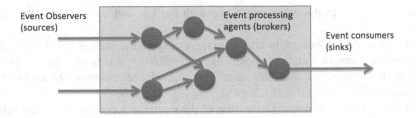

Fig. 4 Event processing framework

2. **Expression of Intent**—the mechanisms through which the user specifies the rules or the relationships between the source events and the sink events.
3. **Performance/Scalability**—since there are a large number of events to be consumed, the performance of event processing is critical to ensure that events are not dropped; hence, low latency of operations is critical.
4. **Fault tolerance**—relates to the aspect that there may be missing data in the input due to network delays and also to the fact that some of the nodes processing the events may go down. The framework must be able to handle all such cases.

We examine each of these requirements in detail in the following sections.

4 Data Model/Semantics

The data model provides a mechanism for the system to model various types of events that are being processed by the system. This can be separated into two parts —an input data model and a data model to express the relationship between various events. The former is described here in this subsection; the topic of the latter is treated in greater detail in separate section.

4.1 Input Data Model

Typically, the data model of the input event is kept fairly simple and the events are expressed mainly as a multi-valued tuple of the type

$$< ts, source - id, key, value >$$

where
ts   refers to the timestamp when the event was generated
source-id   refers to the source node that generated the event
Key   refers to the key associated with event

Value   refers to the value associated with the key and the source at the time
         instant *ts*

For example, in our multi-tank cleaner example, the source event from the
temperature sensor could look like < *2388301, cleaner1, temperature,
45.2* > which indicates that the value read from the temperature sensor is 45.2
degrees and is associated with cleaner1.

This representation permits extension of the model to scenarios where the value
need not just be a single reading. For example, the value could even be an image
file or a URL. This model also ensures that the system is able to take into account
scenarios where only a single value is associated with a point in time and hence
eliminates redundancy from the network.

5 Expression of Intent

The event processing nodes or brokers will have to ingest the incoming events and
process them according to the business rules setup by the user. Various approaches
have been taken to express the user intent to process the incoming events such as
using a query language, providing a programming paradigm or ability to run data
analysis algorithms on the events.

1. **Stream Query Languages and Algebra**—In this approach, a query language is
   defined that operates on a stream of events. The Telegraph CQ project [10]
   defines a continuous query language over PostgreSQL and introduces operators
   to select events from a stream using a sliding window. The STREAM project [26]
   extends SQL to processing an event stream by introducing a sliding window
   specification and an optional sampling clause. For example, it was possible to
   make queries that refer to the previous ten entries. Cayuga [13] also uses an SQL
   like language to specify the queries.

   The Aurora project [2] defines a stream query algebra that permits streams to be
   split, joined, and aggregated. Aurora is capable of working with either streams
   of data or data from a store. The SASE event language is an alternative approach
   that does not draw on SQL [38], but defines events of interest in a given time
   window that match a given pattern.

   Amit [3] uses an XML-based approach to define a situation definition language.
   A situation is a condition within which an event becomes significant and can be
   defined over a lifespan. Borealis [1] provides a mechanism to reapply queries on
   modified data; this might occur when operators realize that a correction is
   required to the stream and hence need to recompute the events.
2. **Graph-based approaches**—In these approaches, the problem of event processing is considered as a network of a loop-free graphs and the user is allowed
   to specify their intent using the graph. Aurora [2] and the SPADE project [18]
   are examples of such systems.

Apache Storm [7, 24] models the processing as a *topology*. The input streams are modeled as *spouts* and the processing nodes are modeled as *bolts*. Twitter uses storm to process incoming tweets [6].

3. **Programming-based approaches**—Yahoo's S4 stream processing engine [27] is inspired by Hadoop and process event tuples across a distributed set of processing elements (PE); each PE defines a *processEvent()* method that is used to process a particular event and in case another event is generated, it invokes the *output()* method.

   The map-update programming paradigm [23] is influenced by the MapReduce model and seeks to ease the task of event processing by breaking down the processing of events into the Map and the Update step and process them over a distributed set of systems. Since events are instantaneous, any state information between events is summarized into a context parameter to the *Update* step known as the *slate*.

4. **Advanced analytics**—Most of the complex event processing frameworks focus on providing an infrastructure to handle large number of events. Additional processing to derive insight from the streams must be built on top [28]. Samoa [25] provides support for a pluggable data mining and machine learning library built over S4 and Storm.

For our manufacturing intelligence application, we would want to specify all queries to include all pressure events within a time period to reach a peak value in the permissible range.

6 Performance/Scalability

The event processing system must be capable of handling events with a low latency. Twitter, for example, must be capable of processing 100000 tweets a minute [14]. In order to process a large number of events, a complex event system must [34]

1. **Avoid storage operations in the critical path**: Storage operations in the critical path lead to additional latency in processing. Aurora [2] avoids all store operations; instead store is performed optionally. This is the approach taken by Apache Storm [7].

2. **Distribute operations**: Gain performance through distribution of workload across multiple nodes. The system must be capable of responding to additional workload by scaling across multiple nodes [7]. The Medusa project provides for distributed naming and catalogs to provide for distributed operations [9]. It also provides mechanism for multiplexing event streams on the TCP layer. Given that events are transferred between processing nodes/brokers that are distributed, the queue sizes in between the nodes are critical to performance. The STREAM project provides a mechanism to optimize the internode queue sizes [26].

3. **Optimize query processing and scheduling**: Queries need to be scheduled on the processing elements to ensure that QoS guarantees are met. The STREAM project [26] provides a structure to the query plans, thus enabling query plan sharing and optimization. It further exploits constraints on data streams to reduce memory overhead during computation.

7 Fault Tolerance

As streams of events are being processed by a distributed set of nodes, there is possibility that either some of the events may not reach or may be delayed in reaching the processing node. Further, it is also possible that some of the nodes that comprise the distributed system may fail. These are handled as

1. **Delayed arrival of events**: Most complex event processing systems work on a sliding window waiting for events [2]. D-stream or discretized streams [39] is alternative approach that processes discretized portions of the input stream to account for lag in the input data. The computation itself is then performed using Spark [40] that uses resilient distributed datasets to ensure fault tolerance.
2. **Failing nodes**: In Muppet [23], each node keeps track of all other nodes in the system. When a node is unable to send data to another node, it assumes that node is not functional and reports failure to the master node. The master node in turn broadcasts the failure status to all the nodes that then update their list of functional nodes. The work is then taken over by another functional node in the system.

Having briefly discussed various features that must constitute a complex event processing system, we delve in more depth in the modeling of complex events.

8 Modeling Complex Events

In this section, we describe some of the important methods that have been proposed for modeling complex events. This would enable us to model the relationship amongst various events and the parameters of interest; for example, in the manufacturing intelligence solution, this would allow us to express the relationship between various sensors and data values read from log files to the quality of the product. The first question is to enumerate the aspects of an event that are important. Next, we survey some of the different approaches that have been proposed for modeling events.

8.1 Event Aspects

Generally, there are many attributes associated with an event. For a complete description of an event, it is necessary to store these attributes. Events commonly occur to one or more objects, which are considered participants in the event. Normally, there is also a context associated with events. This context may be both spatial and temporal. The context need not be absolute—e.g., the temporal context may be a specified time interval after a previous event. The objects participating in an event may have a relationship—e.g., in a cricket game, there will be a batsman and a bowler. Events may also have a hierarchical structure and be nested inside one another.

Figures 5 and 6 [37] list many of the important aspects of events. These include

- The *temporal aspect* of the event, i.e., the time interval over which the event occurred. The time may be the *physical* or absolute time, *relative* to other events or *logical* (domain-specific time) such as the frame number in a video.
- The *causal* aspect of events which links events in a chain of cause and effect is important when analyzing the reasons for occurrence of an event. For example, in studying a sports video, the causal linkages between different events in the game could be important.
- The *spatial* aspect of the event encapsulates where the event occurred. As in the temporal aspect, the spatial aspect may be the *physical* location, or *relative* to other locations, or *logical* (e.g., the home of an individual).
- In multimedia systems, the *experiential* aspects may be important. For example, in instant replay in sports videos, the analyst is seeking to replay the event in order to gain more insight. The media associated with the event is then important.
- The *informational* aspect of the event consists of data about event such as the objects participating in the event, and the type of event (e.g., runs scored in a cricket game).
- Events may be made up of subevents, and these are captured in the *structural* aspects of an event.

Fig. 5 Design aspects of complex event processing systems

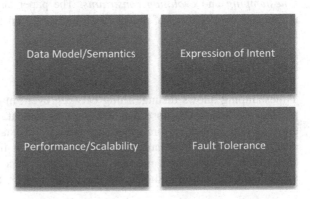

Data Model/Semantics   Expression of Intent

Performance/Scalability   Fault Tolerance

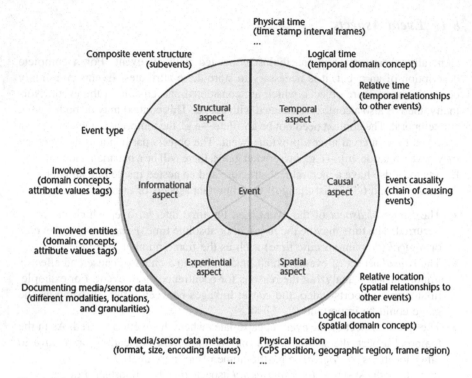

Fig. 6 Aspects of an event

8.2 Categories of Temporal Relationships

In the discussion of event attributes above, we have listed the common types of temporal attributes. Due to having temporal attributes, multiple events can also be linked together in temporal relationships, such as one event occurring before another. Many relationships proposed in different event models were surveyed in [8] where it was shown that these relationships can be divided into two categories—*timestamping* and *evolution constraints*. The paper also contains a discussion of mathematical properties of models with these relationships.

8.2.1 Timestamping

Timestamping allows distinguishing between event model elements that are independent of time, and those that are time dependent. Elements to which timestamping does not apply are called *atemporal*. Timestamping could be applied to entities, attributes, or relationships. An example of timestamping applied to an entity is the lifetime of a machine in a factory. Since the salary of an employee varies with time, timestamping can be applied to an attribute such as employee salary. An example of a timestamped relationship is an employee–manager relationship in a company, which is valid during the time a particular employee has a particular manager.

8.2.2 Evolution Constraints

Evolution constraints govern the evolution of the system being modeled and specify the permissible changes in elements of the model. By specifying evolution constraints, the integrity of the database can be maintained. Some of the important changes include

- *Object Migration*: an object of one class (e.g., employee) becomes an object of another class (e.g., manager). However, migration to director class may be disallowed.
- *Status Class*: here, a timestamp is associated with the object migration, so that during its lifecycle, the object may be said to be in different statuses at different times.
- *Generation Relationship*: an object of one class may generate an object of a possibly different class. An example is the situation where a manager in a business generates a project.

8.3 Survey of Event Models

In the following section, we survey some of the different models that have been proposed by different researchers. These models and their relationships are as follows:

- **Extensions to Relational Models**: Relational databases are widely used as storage systems. Hence, the first model described is the *TimeER* model, which extends the widely used extended entity relationship model for modeling events.
- **Graph Models**: It can be seen that the relationship between different aspects of an event can naturally be modeled by a graph. Therefore, the next model considered is E^*, which proposes a graph-based model of events based upon RDF.
- **Extended Graph Models**: Graph models such as E^* are designed for describing objective events such as events in a surveillance video or in a computer system. Generally, the models are also domain-specific. These models have been extended in different ways by various researchers. Examples include

  - **F**: The *F* model tries to simplify the construction of graph models for events by providing templates for simple events. These simple events can be combined to define complex events.

- **SEM**: The *SEM* model tries to account for historical or fictional events where there can be differing perspectives and interpretations of the event. The interpretation is explicitly modeled in the event model.
- **EventWeb**: The *EventWeb* model is a two-level model where the top level is independent of the domain being modeled, and the lower level is domain-dependent.

8.3.1 Entity Relationship-Based Model for Events

The extended entity relationship model (EER) is a well-known technique for modeling data in a domain. Subsequently, the EER model can automatically be converted into a relational schema [15]. The ER model was extended to capture temporal aspects in [19]. Subsequently, methods for imposing temporal constraints were developed in [11]. The same technique can be used to map temporal ER constraints to NoSQL databases as well. Due to the importance of relational and NoSQL databases, we describe these techniques below. The discussion is based upon the approach in [11].

8.3.2 TimeER Model

The TimeER model extends the EER model by adding timestamping information. Four types of temporal timestamping information are supported. These are

- *Valid time:* the time interval during which a fact is true—for example, the time when an employee is a member of a department.
- *Lifespan:* the time during which an entity exists in the model (not the database). For example, if the database models events in a company, the lifespan of an employee could be the time the employee is in the company.
- *User-defined time:* these are not generic, like the above two times, but depend upon the semantics of the entity. For example, birthdays could be associated with an employee, but not a department.
- *Transaction time:* the time when a fact is in a database.

The concepts behind the TimeER model can be illustrated in Fig. 7, which describes part of the model needed for the manufacturing use case described in Sect. 1. Assume that enough data about the manufacturing process should be captured so that it is possible to find patterns of irregularity in the process that leads to defects. Each *Wafer* is identified by a *WaferId* and can have a *Defect*.[1] The *WaferId* has a lifespan timestamp indicating the time the wafer was in the system. The *Defect* also has a lifespan timestamp indicating when the defect was detected,

---

[1]For simplicity in modelling, assume that each wafer can have at most one defect. The model can be easily extended to multiple defects.

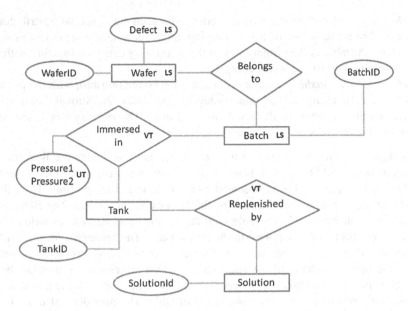

Fig. 7 TimeER model for the manufacturing process

and which may be of use in determining the cause. These timestamps are indicated by the annotation **LS** inside the corresponding boxes. The *Wafer* belongs to a *Batch* which has a *BatchId*. The *Batch* has a lifespan, but the *Belongs to* relationship has been modeled as a permanent relationship, in the same way that a parent–child relation can be considered to be permanent. The *Batch* can be immersed in a *Tank* which has a *TankId*. Since the *Immersed in* relationship holds only during a certain time interval, it has a valid time timestamp, indicated by the **VT** annotation. During the immersion, the wafer is sprayed with a pressure that varies with time. This is represented by sampling the pressure at two different times during the spraying process. This is modeled as the attributes *Pressure1* and *Pressure2* of the *Immersed in* relationship. Each attribute has a user-defined timestamp indicated by the **UT** annotation. Finally, the tank also has to be replenished periodically. This is indicated by the *Replenished by* relationship which, similar to the *Immersed in* relationship, has a valid time timestamp.

8.3.3 Mapping the TimeER Model to a Schema

The above ER model can be mapped into a relational schema. A detailed description can be found in [15]. However, we outline the basics of the process below.

- *Mapping entities:* a table is created for each entity. The entity attributes become the fields of the table, and one of the attributes is selected as the primary key.

- *Mapping weak entity sets:* a table is created for each entity set, with attributes as keys. The primary key of the identifying entity set is also added. The primary key for the table is the primary key of the identifying entity set together with the identifying key of the weak entity set.
- *Mapping relationships:* a table is created for each relationship, with the primary key of each entity in the relationship as the fields. Additional fields could contain the attributes of the relationship (if any). The primary key is the composite primary key of all participating entities.

A *Wafer table* has to be created for wafers. The table will have the *WaferId* and *Defect* as fields. Additionally, it is necessary to store the lifespan for the *Wafer* and the *Defect*. Lifespans can be represented by a start time and an end time. Thus, there would be start times and end times for the *Wafer* and the *Defect*. The *Belongs to* relationship can be stored as a table with the *WaferId* and *BatchId* as fields, since these are the entities participating in the relationship. The *Immersed in* relationship can be stored as a table with *BatchId, TankId, Pressure1,* and *Pressure2* as fields. The valid time timestamp of the relationship, which represents the time the batch was immersed in the tank, can be represented by adding the start and end of the immersion times as fields to the *Immersed in* table. The user-defined time timestamps for *Pressure1* and *Pressure 2* can be stored as additional fields which record the times at which the pressure was recorded. The rest of the TimeER diagram can be similarly converted into a relational schema.

8.3.4 Adding Temporal Constraints to TimeER

Temporal constraints provide constraints on the evolution of the system. Some of the temporal constraints from [11] are listed below. For full details of the constraints, we refer the reader to the original paper. Temporal constraints can apply to keys, attributes, or superclass/subclass relationships as below.

- *Time-invariant key:* constrains the key for the entity to be unchanging during the valid time of the entity.
- *Temporal key:* specifies the constraint that if two entities have the same key, they are the same entity. The difference between this and the time-invariant key is that a temporal key may change with time, but at any given time uniquely references only one entity; whereas a time-invariant key does not change with time.
- *Time-invariant attribute:* constrains the attribute to not change with time.
- *Temporally disjoint subclass/superclass:* constrains an entity to be part of only one subclass.
- *Time-invariant subclass/superclass:* specifies the constraint that the entity's subclass membership cannot change with time.

Figure 8 shows how some of the above constraints can be applied to the TimeER diagram for the manufacturing process. The notation **TI** has been added to the

WaferId, BatchId, SolutionId, and *TankId.* This indicates that those fields are time-invariant keys. Note that in the current model, the *SolutionId* is time-invariant, so keep track only of the solution type. If it is desired to keep track of the solution batch number (i.e., it is assumed that different batches of the solution are used for spraying) then the *SolutionId* could be modeled as a temporal key. These constraints can be enforced by the database.

The TimeER diagram has been modified slightly to enforce the constraint that when the solution in a tank is being refreshed, no batch can be immersed in the tank. The *Tank* superclass has been split into two subclasses—*Replenishing Tanks* and *Replenished Tanks. Replenishing Tanks* are those tanks where the solution is being replenished. *Batches* can only have a *Immersed in* relationship with *Replenished Tanks,* not *Replenishing Tanks.* The notation $d(T)$ indicates that the set of *Replenished Tanks* and *Replenishing Tanks* have a temporally disjoint subclass/superclass relationship, i.e., that batches cannot be immersed in tanks where the solution is being replenished. This constraint can again be enforced by the database.

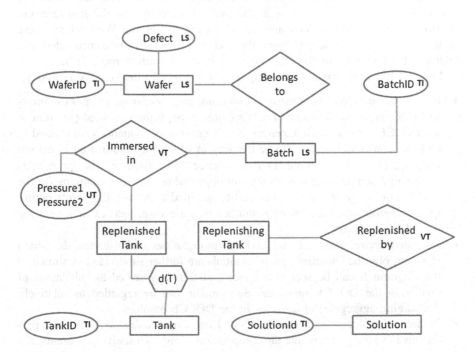

Fig. 8 TimeER model for the manufacturing process with temporal constraints

8.4 Graph-Based Models for Events

The example given above maps events into relational databases. However, since relationships are extremely important in event modeling, many researchers have proposed graph models of events. Graph models of events are useful when modeling events with many relationships, and where the relationships are not known in advance.

In the following we describe E^*, a graph model of events proposed in [20]. This is an extension of the E model proposed in [36]. The temporal aspects of E^* uses the DOLCE ontology defined in [17]. The spatial aspects of E^* are derived using the approach proposed in [29] and is described in a later subsection. The following subsection describes DOLCE. The temporal aspects of E^* are described subsequently.

8.4.1 DOLCE Ontology

The DOLCE ontology is intended to reflect the categories that underlie natural language, and is intended to have a cognitive bias. The need for DOLCE arises from the need for a foundational ontology that can serve as a basis for negotiating meaning between different specialized ontologies that may be developed for extracting semantics from objects in the Web. It is likely that the structure and relationship underlying videos and other documents on the Web reflect these underlying cognitive concepts. Since these concepts are already incorporated into DOLCE, the DOLCE ontology can be a useful tool for this purpose [17].

Figure 9 illustrates many of the important categories in the DOLCE ontology.

- Entities are divided into *Abstract, Endurant,* and *Perdurant* categories. Informally, it can be stated that endurants are objects or attributes (called *Qualities* in the DOLCE ontology) while perdurants are occurrences. Qualities are classed as endurants since this appears to be the way in which they are treated in natural language [17]. For example, in the sentence "The machine's temperature is increasing," temperature is treated as an object whose value is changing (in the DOLCE ontology, the value of a quality is called a *qualia*). Endurants participate in perdurants, e.g., in the manufacturing use case, batches of wafers are dipped in a solution.
- As stated above, endurants are divided into qualities and *substantials,* which represent objects. Qualities and substantials are further subdivided as shown in the diagram. It can be seen that space and time are treated as subclasses of quality in the DOLCE ontology. Substantials can be regarded as relatively unchanging aggregates of qualities in the DOLCE ontology.
- Perdurants or occurrences can be divided into *events* and *statives*. Stative perdurants have the property that the composition of any two stative perdurants of a

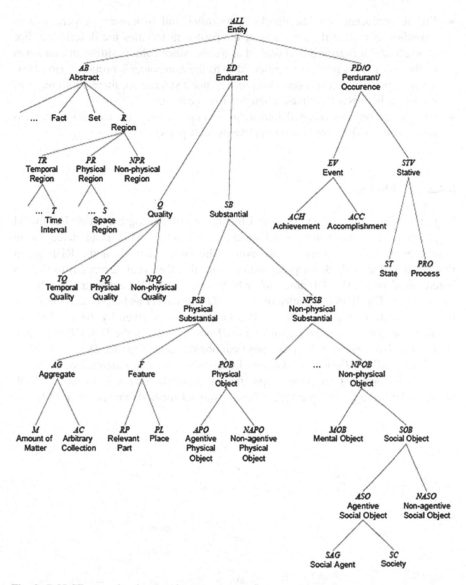

Fig. 9 DOLCE upper level categories

particular type *s* is also a perdurant of type *s*. *John is sitting* is an example of a stative perdurant because the composition of any two (consecutive) perdurants of type *John is sitting* is another perdurant of type *sitting*. *Climbing Mount Everest*, on the other hand, is not a stative perdurant but an eventive perdurant, since the composition of any two perdurants of type *climbing Mount Everest* is not a perdurant of type *climbing Mount Everest*.

- Stative perdurants can be divided into *states* and *processes*. A perdurant is classified as a state if every part of the perdurant satisfies the description. For example, *John is sitting* is classified as a state, since John is sitting in every part of the event. This property is referred to as the *homeomeric* property. However, *John is speaking* is classified as a process, not a state, since there could be short pauses in between the times when John is speaking.
- Events can be subclassified into *achievements* (if they are atomic) or *accomplishments* (if they are compound events with parts).

8.4.2 E* Model

Figure 10 shows part of an *E** model for the manufacturing use case. The model fragment models the batching of wafers into batches. This is modeled as an achievement, since it is an atomic event. The node Ach::01 in the RDF graph denotes this. The *makeBatch* predicate indicates that the event *Ach::01* is to make a batch out of the blank RDF note _wb, which is an ordered list consisting of a *Wafer* and a *Batch*. The *Wafer* has attributes *WaferId* and *Defect* as before, and the *Batch* has an attribute *BatchId*. The time taken for the event is given by the *timeInterval* predicate, which has a *startTime* and an *endTime*. In terms of the DOLCE ontology, *Wafer* and *Batch* are non-agentive physical objects, and *WaferId, Defect, BatchId, startTime,* and *endTime* are qualities. In terms of the implementation, since *E** model is in terms of an RDF graph, the simplest alternative is to use an RDF database for storage and querying. More advanced methods are given in [20].

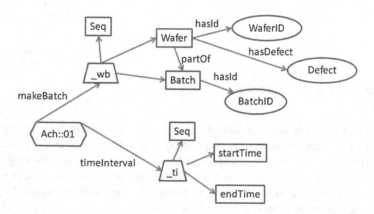

Fig. 10 RDF graph for E* model for the manufacturing process

8.4.3 F

An alternative event model based upon composition of fundamental event patterns is proposed in [31]. The basic idea in *F* is that complex events can be composed by combining the fundamental event patterns. An RDF/OWL implementation of F is available at [30]. The fundamental event patterns of F are

- *Participation:* indicating that an object is participating in an event
- *Temporal duration of events:* the time interval over which an event occurs
- *Spatial extension of objects:* the space occupied by the event
- *Structural relationships:* three types of structural relationships are modeled

  - *Mereological:* indicating that certain events are subevents of others
  - *Causal:* indicating some events cause others
  - *Correlation:* indicating that there is a correlation between events

- *Documentary support:* the media or other documents in which the events are recorded
- *Event interpretations:* an interesting feature of the *F* model is the ability to model an interpretation of an event, since different observers may disagree about the causal relationships or participants in an event.

Figure 11 shows example *F* patterns as well as their usage. Figure 11 shows the participation pattern. It can be seen that the pattern includes all the components of an object participating in an event, such as the type of object, its qualities, and the type of event and its qualities. Similar patterns are defined for all the fundamental patterns described above. To model an event, it is simply necessary to stitch the required patterns together, omitting any null objects or qualities. Figure 12 shows part of the model of the situation where an electric pole has snapped, and has led to

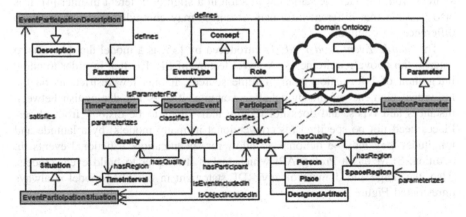

Fig. 11 F participation pattern

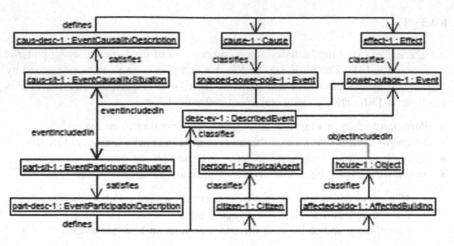

Fig. 12 Combination of F patterns

a power outage. The bottom part of Fig. 12 is derived from the participation pattern, and lists the objects participating in the event—the house and a citizen. The top part is drawn from the causality pattern and shows the snapped telephone pole as the cause of the power outage.

8.4.4 The SEM Model

The models described earlier in this section try to specify precisely the semantics of the various entities in the model. However, in the World Wide Web, it is likely that the information sources would be inconsistent, and that different information sources would model the same information in a slightly different manner [4]. It is also possible that different information sources may contradict each other due to differences in opinion.

The *Simple Event Model* (*SEM*) introduced by [35] is a model that attempts to cater to the above requirements. A novel feature of SEM is that in order to allow integration of information from multiple sources, it puts few restrictions on the individual elements of the model. For example, it does not distinguish between instances and types, and does not require individuals to be disjoint from places. Places need not be specified (as conventional in many models) by a latitude and longitude; they may be fictional, to allow for specification of fictional events. In addition, SEM has a Prolog API which makes it simpler to build event models. Also, the range and domain of any RDF statement in the SEM model is always unrestricted Figure 13.

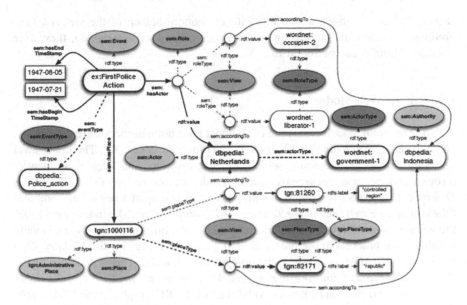

Fig. 13 SEM model for historical statement

There are four core classes in the SEM model. As stated earlier, these need not be disjoint, so that an actor could also be a place. The core classes are

- *sem:Event:* the class that models events
- *sem:Actor:* the entities involved in the event
- *sem:Place:* the place where the event occurred
- *sem:Time:* the time when the event occurred

Additionally, the SEM model also allows specification of property constraints, to capture the constraints present in the event. The role of constraints will be described by considering the SEM model for the historical statement "The Dutch launched the first police action in the Dutch East Indies in 1947; the Dutch presented themselves as liberators but were seen as occupiers by the Indonesian people" [35]. It can be noted that the statement expresses two different points of view—the Dutch, who saw themselves as liberators, and the Indonesians, who saw the Dutch as occupiers. These opposing views are modeled by the blank RDF node that is the object of the *sem:hasActor* predicate (in the top center of Fig. 9). This node points to two different *sem:roleType* nodes. In the top branch, *sem:roleType* leads to an *occupier* value, which is also connected by an *sem:accordingTo* predicate to a *dbpedia: Indonesia*. This sub-graph captures the viewpoint that according to the Indonesians, the role of the Dutch was that of occupiers. The *dbpedia:Indonesia* value is of type *sem:Authority,* which designates a source of data that may conflict with a different source. A similar sub-graph starting with the bottom branch of the *sem:roleType* predicate indicates that according to the Dutch (*dbpedia:Netherlands*), they were liberators. Details can be found in [35].

8.4.5 EventWeb Model

A two-level event model, consisting of an upper level domain-independent ontology, and a lower level domain-specific model, was proposed in [29, 32]. The upper level ontology contains *continuants,* which are time-invariant and keep their identity through change; and *occurrents,* which model events or processes. The class *Dynamic_Entity* models continuants that have dynamic spatial behavior. Temporal relationships are indicated using *Temporal RDF* proposed in [21]. In temporal RDF, the validity of an RDF tuple is indicated by adding the time range in which it is valid within square brackets to the predicate. For example, in Fig. 14, the *Military_Unit* participates in a *Military_Event* in the time intervals t_s and t_e. This is indicated by the RDF tuple (*Military_Unit, participates_in, Military_Event*) $[t_s, t_e]$ where the quantity in square brackets indicates the valid time of the RDF tuple. In the RDF graph, the notation $[t_s, t_e]$ after the *participates_in* predicate similarly indicates the valid time of the predicate. Spatial entities and relationships are based upon a GIS ontology, and contains objects for specifying geographical regions. Relationships available in the GIS ontology include concepts such as *inside*.

It can therefore be seen that the EventWeb lower level model consists of three parts—(i) a temporal model based upon temporal RDF, (ii) a spatial model based upon a GIS ontology, and (iii) a domain-dependent model (called a *theme*). Entities

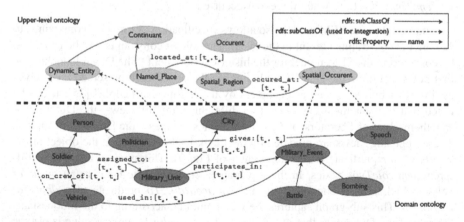

Fig. 14 EventWeb event model

in these three lower level models are mapped to corresponding entities in the upper level ontological model. It can be seen that the EventWeb model is therefore easily generalized to different domains.

9 Conclusions

We have outlined an example of a manufacturing intelligence process and illustrated how modeling it as a complex event model can lead to more structured approach toward monitoring and identifying potential issues in the process. Such formal approaches can also help in identifying root causes and potential impact of various parameters on the manufacturing process.

References

1. Abadi DJ, Ahmad Y, Balazinska M, Cetintemel U, Cherniack M, Hwang J-H, Lindner W et al (2005) The design of the borealis stream processing engine. In CIDR 5:277–289
2. Abadi DJ, Carney D, Çetintemel U, Cherniack M, Convey C, Lee S, Stonebraker M, Tatbul N, Zdonik S (2003) Aurora: a new model and architecture for data stream management. VLDB J: Int J Very Large Data Bases 12(2):120–139
3. Adi A, Etzion O (2004) Amit-the situation manager. VLDB J: Int J Very Large Data Bases 13 (2):177–203
4. Allemang D, Hendler J (2008) Semantic web for the working ontologist. Morgan Kaufman
5. Ananthanarayanan R, Basker V, Das S, Gupta A, Jiang H, Qiu T, Reznichenko A, Ryabkov D, Singh M, Venkataraman S (2013) Photon: Fault-tolerant and scalable joining of continuous data streams. In: Proceedings of the 2013 ACM SIGMOD international conference on management of data, pp 577–588. ACM
6. Aniello L, Baldoni R, Querzoni L (2013) Adaptive online scheduling in storm. In: Proceedings of the 7th ACM international conference on distributed event-based systems, pp 207–218. ACM
7. Apache Storm. https://storm.apache.org
8. Artale A, Franconi E (2009) Foundations of temporal conceptual data models. In: Conceptual modelling: foundations and applications, pp 10–35. Springer, Berlin
9. Cetintemel U (2003) The aurora and medusa projects. Data Eng 51(3)
10. Chandrasekaran S, Cooper O, Deshpande A, Franklin MJ, Hellerstein JM, Hong W, Krishnamurthy S, Madden SR, Reiss F, Shah MA (2003) TelegraphCQ: continuous dataflow processing. In: Proceedings of the 2003 ACM SIGMOD international conference on Management of data, pp 668–668. ACM
11. Combi C, Degani S, Jensen CS (2008) Capturing temporal constraints in temporal ER models. In: Conceptual Modeling-ER 2008. Springer, Berlin, pp 397–411
12. Cugola Gianpaolo, Margara Alessandro (2012) Processing flows of information: from data stream to complex event processing. ACM Comput Surv (CSUR) 44(3):15
13. Demers AJ, Gehrke J, Panda B, Riedewald M, Sharma V, White WM (2007) Cayuga: a general purpose event monitoring system. CIDR 7:412–422
14. Domo. https://www.domo.com/learn/infographic-data-never-sleeps
15. Elmasri R, Navathe SB (2014) Fundamentals of database systems. Pearson,

16. Eyers D, Freudenreich T, Margara A, Frischbier S, Pietzuch P, Eugster P (2012) Living in the present: on-the-fly information processing in scalable web architectures. In: Proceedings of the 2nd international workshop on cloud computing platforms, p 6. ACM
17. Gangemi A, Guarino N, Masolo C, Oltramari A, Schneider L (2012) Sweetening ontologies with DOLCE. In: Knowledge engineering and knowledge management: Ontologies and the semantic Web. Springer, Berlin, pp 166–181
18. Gedik B, Andrade H, Wu KL, Yu PS, Doo M (2008) SPADE: the system S declarative stream processing engine. In: Proceedings of the 2008 ACM SIGMOD international conference on Management of data. ACM, pp 1123–1134
19. Gregersen H (2006) The formal semantics of the timeER model. In: Proceedings of the 3rd Asia-Pacific conference on Conceptual modelling, vol 53. Australian Computer Society, Inc., pp 35–44
20. Gupta A, Jain R (2011) Managing event information: Modeling, retrieval, and applications. Synth Lect Data Manag 3(4):1–141
21. Gutierrez C, Hurtado C, Vaisman A (2005) Temporal rdf. In: The semantic web: research and applications. Springer, Berlin, pp 93–107
22. Johnson T, Shkapenyuk V, Hadjieleftheriou M (2015) Data stream warehousing in Tidalrace. CIDR
23. Lam W, Liu L, Prasad STS, Rajaraman A, Vacheri Z, Doan AH (2012) Muppet: MapReduce-style processing of fast data. Proc VLDB Endowment 5(12):1814–1825
24. Leibiusky J, Eisbruch G, Simonassi D (2012) Getting started with storm. O'Reilly Media, Inc.
25. Gianmarco De Francisci M, Bifet A (2015) SAMOA: scalable advanced massive online analysis. J Mach Learn Res 16:149–153
26. Motwani R, Widom J, Arasu A, Babcock B, Babu S, Datar M, Manku G, Olston C, Rosenstein J, Varma R Query processing, resource management, and approximation in a data stream management system. CIDR
27. Neumeyer L, Robbins B, Nair A, Kesari A (2010) S4: distributed stream computing platform. In: 2010 IEEE International conference on data mining workshops (ICDMW), pp 170–177. IEEE
28. Papadimitriou S, Sun J, Faloutsos C (2005) Streaming pattern discovery in multiple time-series. In: Proceedings of the 31st international conference on Very large data bases, pp 697–708. VLDB Endowment
29. Perry M, Hakimpour F, Sheth A (2006) Analyzing theme, space, and time: an ontology-based approach. In: Proceedings of the 14th annual ACM international symposium on Advances in geographic information systems, pp 147–154. ACM
30. Scherp A, Franz T, Saathoff C, Staab S (2015) http://ontologydesignpatterns.org/wiki/Ontology:Event_Model_F. Accessed 4 Sept 2015
31. Scherp A, Franz T, Saathoff C, Staab S (2009) F–a model of events based on the foundational ontology dolce + DnS ultralight. In: Proceedings of the fifth international conference on Knowledge capture, pp 137–144. ACM
32. Sheth A, Perry M (2008) Traveling the semantic web through space, time, and theme. IEEE Internet Comput 12(2):81–86
33. Sony Corporation, Leading semiconductor wafer surface cleaning technologies that support the next generation of semiconductor devices. http://www.sony.net/Products/SC-HP/cx_news_archives/img/pdf/vol_36/featuring36.pdf
34. Stonebraker M, Çetintemel U, Zdonik S (2005) The 8 requirements of real-time stream processing. ACM SIGMOD Rec 34(4):42–47
35. Van Hage WR, Malaisé V, Segers R, Hollink L, Schreiber G (2011) Design and use of the Simple Event Model (SEM). Web Semant: Sci, Serv Agents World Wide Web 9(2):128–136
36. Westermann U, Jain R (2006) E—A generic event model for event-centric multimedia data management in eChronicle applications. In: Proceedings 22nd International conference on data engineering workshops, 2006, pp x106–x106. IEEE
37. Westermann U, Jain R (2007) Toward a common event model for multimedia applications. IEEE Multim 1:19–29

38. Wu E, Diao Y, Rizvi S (2006) High-performance complex event processing over streams. In: Proceedings of the 2006 ACM SIGMOD international conference on Management of data, pp 407–418. ACM
39. Zaharia M, Das T, Li H, Hunter T, Shenker S, Stoica I (2013) Discretized streams: Fault-tolerant streaming computation at scale. In: Proceedings of the Twenty-Fourth ACM symposium on operating systems principles, pp 423–438. ACM
40. Zaharia M, Chowdhury M, Das T, Dave A, Ma J, McCauley M, Franklin MJ, Shenker S, Stoica I (2012) Resilient distributed datasets: A fault-tolerant abstraction for in-memory cluster computing. In: Proceedings of the 9th USENIX conference on networked systems design and implementation. USENIX Association, pp 2–2

Unwanted Traffic Identification in Large-Scale University Networks: A Case Study

Chittaranjan Hota, Pratik Narang and Jagan Mohan Reddy

Abstract To mitigate the malicious impact of P2P traffic on University networks, in this article the authors have proposed the design of payload-oblivious privacy-preserving P2P traffic detectors. The proposed detectors do not rely on payload signatures, and hence, are resilient to P2P client and protocol changes—a phenomenon which is now becoming increasingly frequent with newer, more popular P2P clients/protocols. The article also discusses newer designs to accurately distinguish P2P botnets from benign P2P applications. The datasets gathered from the testbed and other sources range from Gigabytes to Terabytes containing both unstructured and structured data assimilated through running of various applications within the University network. The approaches proposed in this article describe novel ways to handle large amounts of data that is collected at unprecedented scale in authors' University network.

1 Introduction

In recent times, Computer Networks are fast evolving, with the number of users, applications, computing devices, and network protocols increasing exponentially. However, fundamental challenges still remain, such as scalability, security, and management issues. Today, P2P applications draw a significant attention from network researchers. P2P file-sharing software is often a primary source of information leakage, which has become a serious concern for University network users, and other public organizations. To detect and prevent information leakage from an internal node or user's machine, network administrators must deploy appliances that handle these types of applications in an intelligent manner while providing assurance of

C. Hota (✉) · P. Narang · J.M. Reddy
BITS-Pilani Hyderabad Campus, Hyderabad, India
e-mail: hota@hyderabad.bits-pilani.ac.in

P. Narang
e-mail: p2011414@hyderabad.bits-pilani.ac.in

J.M. Reddy
e-mail: p2011011@hyderabad.bits-pilani.ac.in

© Springer India 2016
S. Pyne et al. (eds.), *Big Data Analytics*, DOI 10.1007/978-81-322-3628-3_9

security and privacy to the end users of these applications. BitTorrent, and other P2P file-sharing networks, allow people to share files amongst themselves. Sharing files on ones' computer with unknown users (peers) on a public Internet by having logical identities brings in natural concerns related to security and privacy. P2P applications open up a TCP or UDP port at the network appliance to get connected to the other side of the application (peer). Hence, once a port is opened, the peer is no longer protected from malicious traffic coming through it. Another major security concern while using a P2P application is how to determine the authenticity of the content that is downloaded or accessed. In other words how can you make sure that you have not received malicious content in the form of spyware or a bot. The kind of impact these P2P applications generate on the performance of university campus networks have not been well studied.

As P2P networks can control their own behavior bypassing the scrutiny from a centralized administrator, it is impossible to implement security protections on a deployed P2P network. Moreover, amongst many peers participating in the applications, it is hard to identify who are those peers who are good and who are bad. Malicious users can actively play the role of insiders. In this paper authors have investigated the negative impacts of P2P traffic on network security and developed privacy-preserving P2P detectors to mitigate these negative impacts. In particular, authors have used their University network to collect and mine comprehensive datasets containing a wide variety of applications like Web, P2P, SMTP, Streaming, etc. The dataset collected also contains regular P2P (benign) and malicious P2P traffic from dormitories, and academic VLANs situated across the campus. Authors have evaluated these datasets quantitatively using existing techniques to reveal how P2P traffic affects the accuracy, sensitivity, and detection capabilities of network security appliances.

This case study of identification of unwanted traffic in large-scale University networks is based on the network of the authors' University. This work provides an evaluation of the impact generated by P2P traffic on University networks, and presents various approaches (existing in literature as well as from authors' prior research) which are used to tackle this kind of unwanted traffic in large-scale University networks.

Our University network serves five thousand to six thousand on-campus students along with faculty and guest users every day, providing wired as well as wireless connectivity in the different parts of the University campus. Internet connectivity is provided in the University by leased lines from two Internet Service Providers (ISPs). The University network consists of a three-tier architecture, consisting of two Core switches (one for fall-back), and multiple Distribution switches and Access switches with a 10 Gb backbone fiber. Wireless connectivity is also provided in the academic blocks and other selected areas within the campus by deploying Wi-Fi access points. Access to different Internet-based services hosted by the University is provided in a 'DMZ' (demilitarized zone). The DMZ hosts the Web (HTTP(S)) service, FTP service, and the Institutional email service (SMTP) of the University. The entire architecture is shown in Figs. 1 and 2. Figure 1 gives the network architecture with respect to the network design of the University, wherein two Core switches and their con-

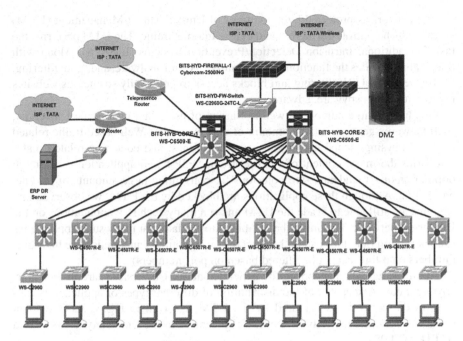

Fig. 1 Network architecture and design of authors' University

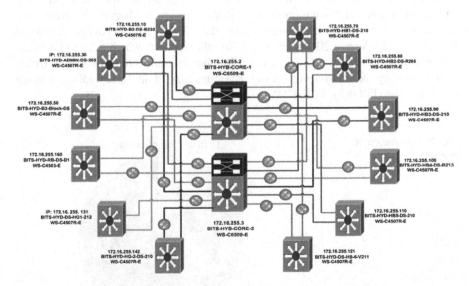

Fig. 2 Network architecture of authors' University—Core and Distribution switches

nections with the ISP line, DMZ, end users, etc., are illustrated. Figure 2 describes the connections of the Core switches with the multiple Distribution switches in the University.

The network is protected by a commercial Unified Threat Management (UTM) system which controls the ingress and egress Internet traffic. The UTM performs the task of traditional Intrusion Detection/Prevention Systems (IDS/IPS). Along with this, it also provides the functionalities of antivirus, anti-malware, and spam filtering. It also performs URL filtering and blocks access to potentially suspicious websites hosting malware, spyware, adware, etc.

Owing to a large number as well as variety of users in a large-scale University, such networks generate a huge amount of data every day. Web-based traffic related to web browsing, social networking, search engines, and education-related material forms the major content of Internet traffic. P2P-based applications—although popular amongst students—are notorious for consuming large amounts of Internet bandwidth. Although these applications are blocked by the UTM, we observed that students are still able to evade the UTM and run many such P2P applications on the University network. This might be attributed to the fact that many such applications use encryption (and thus cannot be detected by signatures) and randomize their port numbers (and thus cannot be filtered based on port numbers).

At an average, the network traffic generated at our University is around one Terabyte per day. A snapshot of the distribution of different types of applications running in the network, as captured by the UTM, is given in Fig. 3. As is evident from the chart, the majority of the traffic generated in the University is web based (HTTP/HTTPS).

The UTM records the web activity of the users of the network, also performs attack prevention and threat monitoring. Figures 4 and 5 give logs of IPS events and Web usage events (respectively) generated by the UTM for a period of one month (20 July 2014 to 20 August 2014). IPS events are those events which refer to intrusion activities identified by the UTM device. Web usage events are those events which are recorded by the UTM device as web activities carried out by several users using protocols like HTTP(S), TCP, FTP, etc.

Fig. 3 Distribution of protocols

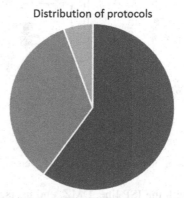

Distribution of protocols

■ https (60%)   ■ http (30%)   ■ others (dns, pop3, imaps) (10%)

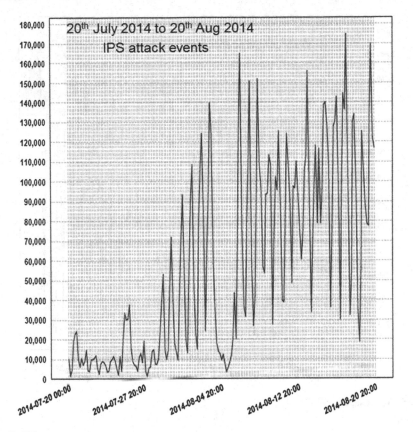

Fig. 4 IPS events

2 Early Challenges in P2P Networks

Any decentralized and distributed computing environment naturally involves many challenges of security, and P2P overlay networks are no exception. In this section, we will review some of the prominent attacks and security threats which were prevalent in P2P networks in their early days (the first decade of the twenty-first century). However, this list is not exhaustive. For a taxonomy of attacks on P2P overlays, we refer the reader to [48, 50].

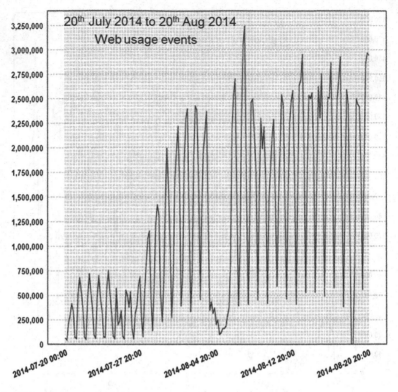

Fig. 5 Web usage events

2.1 Application Level Attacks

2.1.1 Index Poisoning Attacks

P2P file-sharing involves the use of 'indices', which are used by peers to search and find locations of files desired by them. If an attacker inserts massive numbers of bogus records into the index for a set of targeted titles, it is called as an index poisoning attack [25]. When a legitimate peer searches for a targeted title, the index will return bogus results, such as bogus file identifiers, bogus IP addresses, etc.

Key defense against index poisoning is to provide a verifiable identity. This may be done through Public Key Infrastructure (PKI) cryptography, as in [2]. However, PKI requires the presence of a trusted certificate authority, which is not always feasible in a P2P network. Even if a certificate authority is implemented, it will face huge workload due to a large number of peers as well as high peer-churn.

2.1.2 Query Flooding

Flooding of search queries is basic search mechanism employed in many P2P networks. A peer looking for a resource will broadcast its query to its immediate neighbors. If a neighbor does not have the resource, the query is further forwarded to its neighbors. This goes on until the 'hop limit' is reached or the resource is found. Malicious peers can exploit this flooding mechanism to generate a query flooding attack. A malicious user may generate as many useless queries as possible to flood the network. As a result, the resources of the network will be engaged in serving these requests, and benign users may face service degradation.

A solution for preventing query flooding in Gnutella was implemented by putting an upper limit on the number of queries a node can accept from a requesting peer [8]. After this number is reached, all other queries are dropped.

2.1.3 Pollution Attacks

Since peers participate in the P2P network using virtual identities, it is easy for an attacker to spread unusable or harmful content without getting the risk of getting caught. Peers are engaged in downloading chunks of files from different sources. An attacker may host a file x with himself, but replace the original contents with some junk pieces of data or some malware. When a peer, looking for file x, downloads the content from the attacker, he receives 'polluted' content in the form of a corrupted or malicious file. Hash verification and chunk signing are some of the common approaches used against pollution of content in P2P networks [10, 24].

2.1.4 Rational Attacks

P2P networks are based on the principle of peers cooperating to mobilize and share resources. If peers are unwilling to cooperate, the P2P network will fail to survive. A self-interested node, however, may attempt to maximize the benefits it receives from the network while minimizing the contributions it makes. This has been described as a rational attack. A "free-rider" [13] is a term used in P2P systems to describe a peer who utilizes the services and benefit of the network, but does not make any contributions. Since P2P networks are inherently about collaboration and sharing of resources, the services of the network face degradation if most of the users behave in a selfish way and choose to free-ride.

2.2 Identity Assignment Attacks

Peers participate in P2P networks by assuming virtual identities. As a matter of fairness, it is assumed that one physical entity shall own one random virtual identity in the P2P network by which it will participate in network activity. However, a malicious peer may attack this identity assignment principle by creating multiple identities referred to as 'sybil' identities [11]. By positioning its 'sybil' nodes at strategic positions within the network, an attacking peer can try to gain illegitimate control over the network or its part. Sybils can cause multiple damages. By monitoring the traffic flowing through the sybil nodes, the attacker can observe the traffic patterns of other peers whose traffic is flowing through them, and can also attempt to misuse the communication protocol in other ways. Sybil nodes may also be used to forward routing or search requests to other malicious/sybil nodes, thereby disrupting the normal functioning of the network. The attacker can also gain control over certain files shared in the network and can choose to deny access to them or corrupt them.

In DHT-based P2P networks, the nodes having the ID closest to the ID of a particular file are responsible for maintaining information about that file. Hence, an attacker could possibly control the availability of a certain resource if he maps his IDs very close to that resource. Authors in [27] demonstrated the possibility of corrupting the information in the DHT-based Kad network by spreading polluted information through a 'Node insertion' or 'ID mapping' attack. The attacker may carry out this attack to pollute information about keywords, comments or list of sources. Authors in [44] implemented an attack for keywords where the attacker manipulates the ID assignment mechanism of Kad and generates an ID of his own, which lies very close to the targeted file keyword. Hence, when a search is carried out for that particular keyword, it does not yield correct results. Rather, the bogus information planted by the attacker is returned.

A number of approaches have been suggested to defend P2P systems against sybil attacks. One of the early work in this regard is the use of computational puzzles [4]. Authors in [18] evaluate the use of CAPTCHAs in detecting sybils. An approach for detecting sybils using 'psychometric tests' is proposed in [19]. Authors in [49] have argued that a malicious user can create many identities, but only a few relationships of trust. They use this fact to locate disproportionately small 'cuts' in the graph between the sybil nodes and the honest nodes, and attempt to bound the number of identities a malicious user may create.

2.3   Routing Attacks

2.3.1   Route Table Poisoning Attacks

Due to high rate of churn (joining and leaving of peers) in P2P networks, peers have to regularly update their routing tables in order to perform correct routing lookup. Peers create and update their routing tables by consulting other peers. If a malicious node sends false information in its routing table update, it will corrupt the routing table entries of benign nodes, leading to queries being directed to inappropriate nodes or nonexistent nodes.

Different solutions have been developed to counter route table poisoning by imposing certain requirements on the participating peers. In the Pastry network [40], each entry in routing tables is preceded by a correct prefix, which cannot be reproduced by a malicious entity. The CAN network [37] considers the round-trip time in order to favor lower latency paths in routing tables. Authors in [6] propose 'induced' churn as a counter against such attacks.

2.3.2 Eclipse Attacks

Each node in the P2P network maintains overlay links to a set of neighbor nodes, and each node uses these links to perform a lookup from its neighbors. Thus, an attacker may be able to control a significant part of overlay network by controlling a large part of the neighbors of legitimate nodes. This is known as an eclipse attack. Eclipse attacks are *escalated* forms of identity assignment attacks or route table poisoning attacks described above. If an attacker is able to generate a large number of fake identities and place those identities in the overlay network, he could mediate most overlay traffic, and thus *eclipse* legitimate nodes from each other. A pictorial representation of node insertion attacks and sybil attacks being escalated into an eclipse attack is given in Fig. 6.

Authors in [45] describe the eclipsing of search keywords in Kad P2P network. Their experiment explains that an eclipse attack and a sybil attack can be performed quite similarly, except that the Kad ID space covered for an eclipse attack is much smaller. For eclipsing a certain keyword k in a DHT-based network such as Kad, the

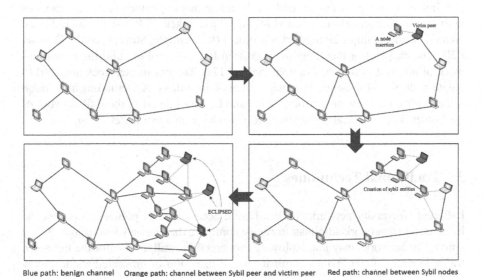

Blue path: benign channel   Orange path: channel between Sybil peer and victim peer   Red path: channel between Sybil nodes

Fig. 6 Node insertion attack and sybil attack escalated to an eclipse attack

authors in [45] choose to position a certain number of sybils as close as possible to the keyword k. Then, the sybil nodes are announced to the benign peers in the network which has the effect of 'poisoning' the regular peers' routing tables for k and attract all the route requests for k to the sybil nodes. Authors in [45] claim through their experiments that as few as eight sybil peers were sufficient to ensure that any request for k terminates on one of the sybil nodes which were strategically planted in the network to lie as close as possible to k. Since all requests for k go via the sybil nodes, the attacker has effectively *eclipsed* the keyword k from all peers in the Kad network.

The developer community of eMule (a P2P application which uses the Kad network) had responded to these security threats by making suitable changes to their applications. For example, the 'change-log' for eMule's version 0.49a released in February 2008 (available at http://www.emule-project.net) states, "Kad will now enforce certain limits when adding new contacts to the routing table: No more than 1 KadNode per IP, 2 similar KadNodes (same bin) from a /24 network and max 10 different KadNodes from a /24 network are allowed". Some countermeasures have also been proposed in literature, such as an optimized routing table and verified routing table [5], and a combination of induced churn and one-hop redirection [35]. Authors in [51] presented a scheme for generating node IDs which requires a user to solve a computational puzzle generated by their network parameters together with time-related information. The authors claim that such an ID generation mechanism makes ID assignment and eclipse attacks computationally infeasible for an attacker, while levying only a small overhead on a regular user.

The implementation of many such countermeasures made large eclipse attacks very difficult or computationally prohibitive for an attacker. But, the emergence of P2P botnets provided a new ground for the adversaries to launch large-scale attacks without the time and effort required for an eclipse attack. P2P botnets can either be created on an existing P2P network's protocol (for example, Storm used the Overnet P2P network [46]) or they may use a custom P2P protocol to build their own network of infected hosts (for example, Sality [12]). The bot master does not need to create multiple sybil identities or perform node insertions. A victim machine, upon infection from a malware or trojan, joins and becomes a part of the P2P network of the botnet. The decentralized architecture also helps in evading detection.

3 Traditional Techniques

Different University networks have different access control policies to access the Internet, and a network administrator must configure the Firewall accordingly. Many University networks may not deploy a commercial firewall to monitor the unknown traffic in each subnet. To accomplish task of detection of unwanted traffic, many open-source tools are available, such as Snort [39], Suricata [9], Bro [29], etc. In this section, we will discuss traditional approaches for P2P traffic detection using port-based, payload-based, and machine learning-based approaches.

3.1 Port-Based Classification

The Internet Assigned Numbers Authority (IANA) has a list of registered port numbers of a known application. Using these port numbers, an application communicates between two IP endpoints over TCP or UDP. IANA has defined some 'well-known' port numbers which are in the range 1–1024. Port-based classification technique is a traditional approach of classifying Internet traffic based on the port number being used by an application [42]. If a network administrator wants to block based on an application using this approach, many false positives (FP) and false negatives (FN) may be generated. This is explained as follows. First, some applications are not registered with IANA (e.g., certain P2P applications). Furthermore, certain P2P applications masquerade their traffic by using well-known port numbers used by web applications (port 80 and 443). 'Torrent' based applications have this property that their initial communication uses port 80 to contact the tracker server. Once the server replies with the peer list, the application uses some random port number to accomplish the rest of the task. In certain cases, if IP layer security is enabled, it is almost impossible to even know the port numbers used.

3.2 Payload-Based Classification

Payload-based detection approach inspects the payload contents of each and every packet. 'Signatures' are extracted for different applications from their TCP/UDP/HTTP payloads [43]. These signatures are employed by detection tools to and improve the accuracy of unwanted traffic detection. This approach is also known as Deep Packet Inspection (DPI). This technique can accurately classify the traffic at wirespeed. Although this technique is highly accurate, it requires significant amount of processing power. Furthermore, this approach proves less accurate on encrypted and proprietary protocols. Furthermore, the widespread commercial deployments of such DPI products has ensued a global debate on the ethical, legal and privacy implications of this practice.

In our initial experiments, we used payload-based identification of P2P traffic using an open-source IDS Snort. To detect P2P traffic, we extracted signatures from multiple P2P applications from network traces obtained from our campus. A sample of these signatures is given in Table 1.

3.3 Classification Based on Machine Learning

The techniques described in previous sections depend upon inspection of the transport/application layer signature and port numbers. These approaches have certain limitations, as mentioned in the previous sections. Later, researchers have devised

Table 1 Application signatures

Application	TCP	UDP
DC++	$Send, $Connect, $Get, $Search, $SR, $Pin, $MyINFO, $MyNick, $Hello, $Quit, $ADCGET, $ADCSND, $Supports, $dcplusplus	–
BitTorrent	GET, /announce, info_hash=, peer_id=, event=, 13 BitTorrent protocol	–
GNUTella	GNUTELLA, GNUTELLA CONNECT, GNUTELLA OK	–
eDonkey	Server 3A eMule	–

new approaches based on statistical characteristics to overcome the limitations of preceding approaches and accurately identify traffic application with low complexity, and ensure users' privacy at the same time [21, 23]. The underlying idea is that the traffic has certain statistical patterns at network layer that are unique for different classes of applications (e.g., packet lengths, inter-arrival time of packets, bytes transferred per second, etc.). To deal with statistical patterns in multidimensional datasets generated from traffic flows, Machine Learning (ML), and pattern recognition techniques were introduced in this field. ML is a powerful technique in the area of data mining which finds useful patterns from the given data. ML techniques are broadly categorized into two learning groups: Supervised learning and Unsupervised learning.

4 Approaches for Identification of Unwanted Traffic

4.1 Dataset

Owing to their large number and variety of users, University-wide networks generate a large amount and variety of traffic. For the purpose of this case study, we obtained network logs from the backbone router of our University. Majority of this data is of web-based traffic. Apart from this, we also captured the data of popular file-sharing P2P applications which are commonly used by on-campus students. DC++ is a popular P2P application used for files sharing on the network LAN. Certain Internet-based applications like BitTorrent, eMule are also popular amongst students. The details of this dataset are given in Table 2.

Apart from benign Internet data, we also gathered data for malicious applications such as botnets. Part of this data was generated in a testbed at our University by running the executables of these botnets in a controlled, virtualized environment. Part of it was obtained from other sources, namely University of Georgia [36], University of

Table 2 Benign dataset details

Application	No. of packets (K)	No. of bytes
HTTP (S)	265	1.93 GB
SMTP/POP3	55	40 MB
FTP	1000	99.6 MB
DC++	1844	20 GB
P2P apps (BitTorrent/eMule)	139	90 MB

Table 3 Malicious dataset details

Botnet name	What it does?	Type/size of data	Source of data
Sality	Infects exe files, disables security software	.pcap file/ 25 MB	Generated on testbed
Storm	Email Spam	.pcap file/ 4.8 GB	University of Georgia
Waledac	Email spam, password stealing	.pcap filc/ 1.1 GB	University of Georgia
Zeus	Steals banking information by MITM key logging and form grabbing	.pcap file/ 1 GB	Univ. of Georgia, Czech Technical University + Generated on testbed
Nugache	Email spam	.pcap file/ 58 MB	University of Texas at Dallas

Texas at Dallas [28] and Czech Technical University [15]. The details of this dataset are given in Table 3.

4.2 Ensemble Learning Approach

One of the promising approaches to identify unknown traffic at University backbone networks is by using ML techniques. Our approach is shown in Fig. 7. In this process we accumulate the packets into flows. We constructed 'flows' using the 5-tuple, i.e., $\langle SourceIP, DestinationIP, Sourceport, Destinationport, Protocol \rangle$.

Ensemble learning techniques combine multiple individual models for better prediction. The performance of our approach using ensemble learning techniques like *Stacking* and *Voting* is shown on Fig. 8. In this section, we first define *Stacking* and *Voting*, which are two of the most well-known ensemble approaches. Then, we discuss our approach for identification of P2P traffic using ensemble learning [38].

- *Stacking*: This technique has three layers as shown in Fig. 8. In the first layer, it takes the training dataset as input. The second layer, the input dataset is fed to the

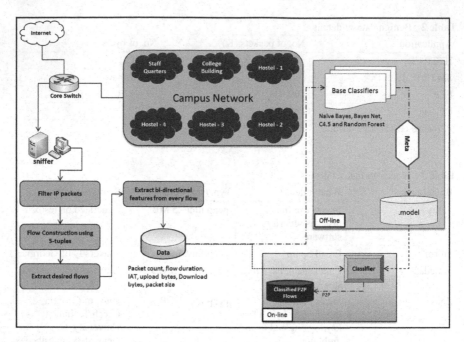

Fig. 7 Flow-based approach

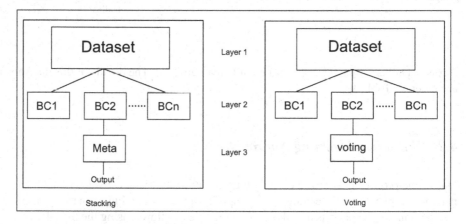

Fig. 8 Ensemble learning

several 'base classifiers' to obtain the individual predictions. In the third layer, a 'meta classifier' is employed to obtained the final prediction.

- *Voting*: This approach is similar to that of *Stacking* till the second layer. The only difference is that there is no meta classifier used in this technique. In the third layer, it combines the all probability estimations from the base classifiers and classifies an instance based on the majority vote.

Table 4 Ensemble classifiers

	Recall			Accuracy		
	CSE	PCA	Full	CSE	PCA	Full
Stacking	99.89	99.6	99.9	99.89	99.66	99.94
Voting	99.7	99	99.7	99.79	99	99.77

Our approach using ensemble learning:

- Step I: We captured the network traffic at our core switch and stored them in SAN.
- Step II: Once desired captures are obtained, we preprocess the traces and filter only IP packets. Then, we obtained desired flows to extract features. We extracted a total of 24 features from each bi-flows (i.e., a pair of unidirectional flows between the same 5-tuple) and stored them into comma separated values (CSV) format.
- Step III: After feature extraction, we labeled each instance according to class as P2P or non-P2P for classification.
- Step IV: We used ensemble learning technique to classify network traffic. In the offline phase, we experimented with Naïve Bayes, Bayesian Networks, decision trees as base classifiers and random forest as decision classifier. In online phase, we classified unknown samples (i.e., no class label is assigned to it) using model built from off-line phase. The experimental results obtained are shown in Table 4 [38]. The table shows Accuracy and Recall of Stacking and Voting for full feature set and reduced feature sets using CSE and PCA. For more details on CSE and PCA, please refer to Sect. 4.5. Ensemble learning was used with Naïve Bayes, Bayesian Networks, Decision trees as base classifiers, and Random forest as the meta classifier.

Our experiments show that Stacking performs better than Voting as number of base classifiers increase. However, *Stacking* suffers from a drawback that it needs more 'build time' than *Voting*.

4.3 Conversation-Based Mechanisms

Five-tuple flow-based categorization has been the *de-facto* standard for Internet traffic analysis. This classical definition relies on port and protocol. Flow-based approaches have seen wide use in problems of Internet traffic classification [21, 22]. However, recent literature [26, 33] has reported that modern malware and botnets are known to randomize their port numbers and switch between TCP and UDP for their communication. Such behavior can allow malware to circumvent traditional security mechanisms which rely on 'flow' definition. By randomizing ports and switching between TCP and UDP, a malicious application creates multiples flows of what is actually a single 'conversation' happening between a pair of nodes. Even many P2P applications like BitTorrent are known to randomize ports in order to circumvent Firewalls which try to block the standard ports used by these applications.

In this section, we discuss a conversation-based approach targeted toward identification of *stealthy* botnets in Internet traffic [33]. Stealthy botnets try to evade standard detection mechanisms by generating little traffic and perform malicious activities in a manner not observable to a network administrator. This may include the Command & Control (C & C) activities of a botnet or other nefarious activities like spamming, Bitcoin mining, etc.

A conversation-based approach is port- and protocol-oblivious. Herein, we begin by extracting five-tuple flows occurring between all pair of nodes seen in network traffic based on a time-out parameter (TIMEGAP). For every pair of nodes, the flows are further *aggregated* in conversations. Multiple flows between two nodes are aggregated into a single conversation given that the last packet of the previous flow and first packet of the next flow occur between a FLOWGAP time period. The FLOWGAP is tunable parameter which is to be decided by a network administrator. Certain botnets may try to pass under the radars of Intrusion Detection Systems/Firewalls by generating very little traffic. A high FLOWGAP can be useful for observing the conversations of such stealthy botnets. Our experiments in [33] indicated that a FLOWGAP value of one hour is suitable for most purposes.

It has been reported that a observing a flow for a larger time window of, say, one hour aids in higher accuracy of malicious traffic detection [36]. However, a large time window will create network traces of huge size. With the increasing size of large-scale networks, parsing such traces create a huge performance overhead. A conversation-based approach aids in this aspect by following a two-tier architecture. In the first step, flows are created from small time windows of, say, ten minutes. Parsing flows of ten minute duration does not create a performance overhead. As a second step, the flows between two nodes are aggregated into a single conversation (based on the FLOWGAP parameter explained above). This approach can be further enhanced by clustering the flows created in the first step, and then creating conversations of flows *within each cluster*. For more details of this approach, we refer the reader to [31].

In order to evade detection by Intrusion Detection Systems, malware and bot-infected machines often exhibit a 'low-lying' behavior. They are stealthy in their activities and do not generate much traffic. However, in order to *not* lose connectivity with their peers, bots engage in regular exchange of messages with each other. For example, we observed in the network traces of the Zeus botnet that the bot regularly sent out as few as one or two packets every half an hour over UDP. Conversation-based approaches harness this 'low volume, high duration' behavior of stealthy botnets. Since conversation-based approaches do not rely on port numbers or Transport layer protocol for the detection of malicious traffic, they cannot be circumvented by bots using these mechanisms.

After creating conversations between a pair of nodes, statistical features are extracted from each conversation. These features are aimed at capturing the 'low volume, high duration' behavior of stealthy botnets. The features extracted are:

1. The duration of the conversation.
2. The number of packets exchanged in the conversation.
3. The volume of data exchanged in the conversation.
4. The median value of the inter-arrival time of packets in that conversation.

These features are used to build classification modules utilizing supervised Machine learning algorithms for separating malicious traffic from benign traffic. We employed three supervised machine learning algorithms—Bayesian Network, J48 Decision trees, and Boosted Reduced error-pruning (REP) trees. The classification results (adopted from [33]) obtained with the data of two malicious applications (namely Storm and Waledac) and two P2P applications (namely uTorrent and eMule) are given in Fig. 10. TP rate represents 'True Positive rate,' and FP rate represents 'False Positive rate.' An outline of this approach is given in Fig. 9.

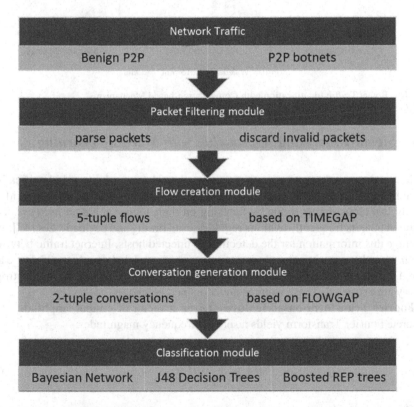

Fig. 9 System overview: conversation-based mechanisms

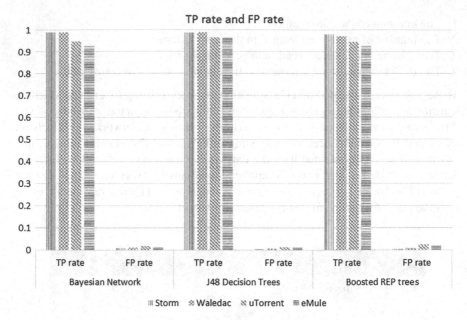

Fig. 10 Botnet Traffic Identification with Conversation-based Mechanisms

4.4 Signal-Processing Approaches with Machine Learning

In this section, we discuss a signal-processing-based approach toward identification of malicious traffic [32]. By utilizing signal-processing approaches, we aim at identifying malicious botnet traffic by focusing on their *timing* and *data patterns*. The traffic of botnets frequently exhibits regularities in timing and payload sizes [47]. We leverage this information for the detection of infected hosts. Internet traffic between a pair of nodes seen in network communication is modeled as a 'signal.' For each signal, several features based on Fourier Transforms (DFTs) and Shannon's Entropy theory are extracted. These features are explained below:

Fourier Transform-based features: Given a vector of n input amplitudes, the Discrete Fourier Transform yields a set of n frequency magnitudes:

$$X[k] = \sum_{n=0}^{N-1} x(n) \cdot e^{\frac{-2j\pi kn}{N}} \tag{1}$$

Here, N is the length of the sequence to be transformed, k is used to denote the frequency domain ordinal, and n is used to represent the time-domain ordinal.

We apply Fourier Transform to the Inter-arrival time of packets in a conversation and to the payload length (in Kbs) per packet in a conversation. In any time series, most of the energy is contained in the first few Fourier Transform coefficients itself. Thus we sort the Fourier Transform values by magnitude, and retain the top

10 Fourier Transform magnitude values and corresponding phase values. We also retain the magnitude and phase of the 'prime wave' obtained by the vector sum of the top ten peaks.

Shannon's Entropy-based feature: Botnets primarily engage in exchange of C & C information. Their communication frequently occurs in small, fixed-size payloads. On the other hand, Internet traffic exhibits much more variation in payload sizes. This can be attributed to the fact that different users of the Internet use different applications. The combination of multiple users and multiple applications leads to greater variety.

We quantify the amount of randomness or entropy in payload variation in each conversation in the form of 'Compression ratio'. We model payload size per packet as a discrete symbol, and compute the expected compression ratio based on Shannon's Entropy theory. The Entropy $H(X)$ is given by:

$$H(X) = -\sum_{x \in X} p(x) \cdot \log_2(p(x)) \tag{2}$$

Benign conversations are expected to have high entropy, which will lead to less compression. On the other hand, botnet conversations are expected to have low entropy, leading to higher compression rates.

We combine our signal-processing-based features with traditional network behavior-based features for each conversation (such as, Average payload size, Number of packets, Median of inter-arrival time between packets, etc.). These set of features were used to build detection models with supervised Machine learning algorithms. For more details, we refer the reader to [32].

4.5 Feature Subset Selection Approaches

Feature subset selection approaches can be utilized to select a subset of features which highly contribute towards the classification, and remove those features which do not provide high contribution toward the classification problem. The number of features used in training Machine learning models has a direct relationship with the 'build time' of the model. By getting rid of irrelevant features, computational complexity of training the models as well as classification can be reduced.

We measure the impact of feature subset selection on three well-known Machine learning algorithms—Naïve Bayes, Bayesian Network and J48 Decision trees. We used three feature subset selection approaches—Correlation-based Feature Selection (CFS) [16], Consistency-based Subset evaluation (CSE) [7] and Principal Component Analysis (PCA) [20]. We compared the build times of each of these approaches with the build time for the full feature set [34]. We observed a pattern in build times for different models, irrespective of the classifier used. Higher number of features *always* led to more build time, while lesser number of features always assured lesser building time for the model. Thus, the full feature set had the maximum build time.

The build time of CFS, CSE, and PCA is in the order of the number of features selected by them. However, we clearly observed that Decision trees, in general, take long time to build. Naïve Bayes classifier was the fastest, followed by Bayesian network.

4.6 Graph Partitioning Approaches for Botnet Traffic Identification

Conventional graph partitioning algorithms—such as community detection algorithms, sybil detection algorithms, etc.—leverage the presence of a bottleneck 'cut' separating a subgraph of interest from the rest of the graph. These approaches are designed for vertex partitioning. This is not applicable for identification of botnets in network traffic [30]. Botnet edges and edges corresponding to legitimate traffic cannot be separated by a bottleneck cut since they can belong to the same node. Thus, identification of botnet traffic is better defined as an edge-partitioning problem from graph-theoretic perspective.

Most of the previous botnet detection approaches leveraging graph theory rely on vertex partitioning approaches. BotTrack [14] attempts to identify highly connected subgraphs of the traffic dispersion graph. It uses the page-rank algorithm to obtain page-rank scores corresponding to 'hub' and 'authority' values. The nodes in the graph are then clustered based on the two dimensional vector of hub and authority value. A well-known community detection algorithm, the Louvain algorithm [3], has been used by Ruehrup et al. [41] and Hang et al. [17] to detect 'botnet communities' in network traffic. Since these approaches rely on vertex-partitioning techniques, they will be unable to categorize whether a node belongs to two (or more) communities (say, one 'botnet' community, and another benign community).

Botyacc [30] is a recent work which uses edge-partitioning approach for P2P botnet detection in large networks. However, neither Botyacc nor most of past research employing graph theory has evaluated the detection of P2P botnets in the presence of benign P2P traffic. Detection of P2P botnets in the presence of benign P2P applications such as uTorrent, eMule, etc. is a challenging task since both these traffic utilize P2P communication topologies. Thus, if a system running a P2P application gets infected by a P2P botnet, such approaches may not be able to correctly categorize its traffic.

We evaluate vertex-partitioning and edge-partitioning approaches for the detection of P2P botnets in the presence of benign P2P traffic in a network. For vertex partitioning, we resort to the well-known approach of Louvain algorithm. For edge-partitioning, we use a recent approach of 'link communities' given by Ahn et al. [1]. We created a graph from network flows of one P2P application eMule and two P2P botnets, Zeus, and Waledac. The data of eMule had one source node engaged in P2P communication with several destination nodes. The data of Zeus contained one source node, and that of Waledac contained three source nodes. We overlapped the eMule node with one of the Waledac nodes. The graph thus obtained contains a P2P botnet and a P2P application *overlapped* on a single node. We then used Louvain algorithm for vertex-partitioning and Link communities algorithm for edge-partitioning. The resultant graphs are given in Fig. 11.

As expected, we see Zeus as a separate community in both the graphs (represented in red in Fig. 11a and in green in Fig. 11b). Louvain algorithm could create two communities for the Waledac-eMule subgraph. In reality, however, this should have been four communities—three for the three source nodes of Waledac, and one for the one source node of eMule.

(a)   **(b)**

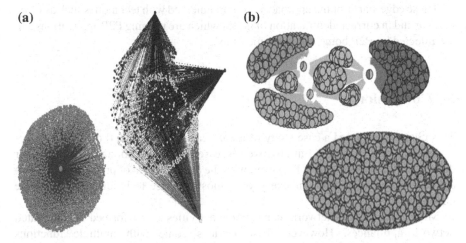

Fig. 11 Vertex-partitioning and Edge-partitioning approaches with P2P botnets and P2P applications

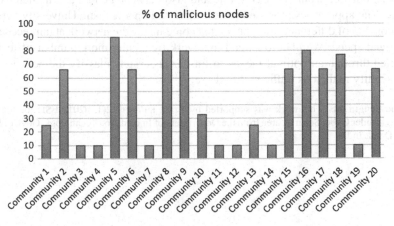

Fig. 12 Sample results obtained using edge-partitioning approach

With the edge-partitioning approach, we obtained a total of 56 communities. Out of these, only four communities were prominent, and others contained only three or four nodes. We have shown only these prominent communities in the Fig. 11b. This approach was successful in creating four *overlapping* communities for the Waledac-eMule subgraph.

We calculate the percentage of malicious edges in different communities attached to a particular node, and then obtain a weighted average of that percentage. In this way we arrive at the *maliciousness* of a particular node. Sample results obtained from this approach are given in Fig. 12, wherein the percentage of malicious nodes detected in a community are indicated.

These edge-partitioning approaches, complemented with techniques such as Boty-acc, can aid in correct identification of nodes which are running P2P applications and are infected by P2P botnets at the same time.

5　Conclusion

This chapter presented a case study of identification of unwanted traffic at a large-scale University network. Initially, we discussed about traditional approaches for detection of unwanted traffic. However, with the advancement of protocols and introduction of newer applications every year, most of these techniques have become obsolete.

Most of the critical network management activities are performed by automated network appliances. However, these devices come with multiple functions that degrade their efficacy. In this paper, we have brought out techniques based on Machine learning approaches to clearly segregate malicious traffic from non-malicious traffic without performing Deep Packet Inspection that would otherwise have been costly. The intelligent approaches proposed in this work clearly separate out P2P from non-P2P traffic and also between benign and malicious P2P traffic. The approaches proposed have been tested on a realistic University network that consists of different types of data that come inside the network at unprecedented scale and speed. We plan to extend this work by adding more number of features selected through an optimal means to further accurately classify the network logs and identify malicious traffic within those.

Acknowledgments This work was supported by Grant number 12(13)/2012-ESD for scientific research under Cyber Security area from the Department of Information Technology, Govt. of India, New Delhi, India.

References

1. Ahn YY, Bagrow JP, Lehmann S (2010) Link communities reveal multiscale complexity in networks. Nature 466(7307):761–764
2. Berket K, Essiari A, Muratas A (2004) Pki-based security for peer-to-peer information sharing. In: Fourth international conference on peer-to-peer computing, 2004. Proceedings. IEEE, pp 45–52
3. Blondel VD, Guillaume JL, Lambiotte R, Lefebvre E (2008) Fast unfolding of communities in large networks. J Stat Mech: Theory Exp (10):P10,008
4. Borisov N (2006) Computational puzzles as sybil defenses. In: Sixth IEEE international conference on peer-to-peer computing, 2006. P2P 2006. IEEE, pp 171–176
5. Castro M, Druschel P, Ganesh A, Rowstron A, Wallach DS (2002) Secure routing for structured peer-to-peer overlay networks. ACM SIGOPS Oper Syst Rev 36(SI):299–314
6. Condie T, Kacholia V, Sank S, Hellerstein JM, Maniatis P (2006) Induced churn as shelter from routing-table poisoning. In: NDSS
7. Dash M, Liu H (2003) Consistency-based search in feature selection. Artif Intell 151(1):155–176
8. Daswani N, Garcia-Molina H (2002) Query-flood dos attacks in gnutella. In: Proceedings of the 9th ACM conference on computer and communications security., CCS '02ACM, New York, NY, USA, pp 181–192
9. Day DJ, Burns BM (2011) A performance analysis of snort and suricata network intrusion detection and prevention engines. In: The Fifth international conference on digital society
10. Dhungel P, Hei X, Ross KW, Saxena N (2007) The pollution attack in p2p live video streaming: measurement results and defenses. In: Proceedings of the 2007 workshop on peer-to-peer streaming and IP-TV. ACM, pp 323–328
11. Douceur JR (2002) The sybil attack. In: Peer-to-peer systems. Springer, pp 251–260
12. Falliere N (2011) Sality: story of a peer-to-peer viral network. Symantec Corporation, Rapport technique
13. Feldman M, Papadimitriou C, Chuang J, Stoica I (2006) Free-riding and whitewashing in peer-to-peer systems. IEEE J Sel Areas Commun 24(5):1010–1019
14. François J, Wang S, State R, Engel T (2011) Bottrack: tracking botnets using netflow and pagerank. In: Proceedings of the 10th International IFIP TC 6 conference on networking—volume Part I, NETWORKING '11. Springer, Berlin, pp 1–14
15. García S, Grill M, Stiborek J, Zunino A (2014) An empirical comparison of botnet detection methods. Comput Secur
16. Hall MA (1999) Correlation-based feature selection for machine learning. PhD thesis, The University of Waikato
17. Hang H, Wei X, Faloutsos M, Eliassi-Rad T (2013) Entelecheia: detecting p2p botnets in their waiting stage. In: IFIP networking conference, 2013. IEEE, USA, pp 1–9
18. Haribabu K, Arora D, Kothari B, Hota C (2010) Detecting sybils in peer-to-peer overlays using neural networks and captchas. In: 2010 International conference on computational intelligence and communication networks (CICN). IEEE, pp 154–161
19. Haribabu K, Hota C, Paul A (2012) Gaur: a method to detect sybil groups in peer-to-peer overlays. Int J Grid Util Comput 3(2):145–156
20. Jolliffe I (2005) Principal component analysis. Wiley Online Library
21. Karagiannis T, Broido A, Faloutsos M et al (2004) Transport layer identification of p2p traffic. In: Proceedings of the 4th ACM SIGCOMM conference on internet measurement. ACM, pp 121–134
22. Karagiannis T, Papagiannaki K, Faloutsos M (2005) Blinc: multilevel traffic classification in the dark. ACM SIGCOMM Comput Commun Rev 35:229–240 (ACM)
23. Li J, Zhang S, Lu Y, Yan J (2008) Real-time p2p traffic identification. In: Global telecommunications conference, 2008., IEEE GLOBECOM 2008. IEEE, USA, pp 1–5

24. Liang J, Kumar R, Xi Y, Ross KW (2005) Pollution in p2p file sharing systems. In: INFOCOM 2005. 24th Annual joint conference of the IEEE computer and communications societies. Proceedings IEEE, vol 2. IEEE, pp 1174–1185
25. Liang J, Naoumov N, Ross KW (2006) The index poisoning attack in p2p file sharing systems. In: INFOCOM. Citeseer, pp 1–12
26. Livadas C, Walsh R, Lapsley D, Strayer WT (2006) Using machine learning techniques to identify botnet traffic. In: 31st IEEE conference on local computer networks, proceedings 2006. IEEE, pp 967–974
27. Locher T, Mysicka D, Schmid S, Wattenhofer R (2010) Poisoning the kad network. In: Distributed computing and networking. Springer, pp 195–206
28. Masud MM, Gao J, Khan L, Han J, Thuraisingham B (2008) Mining concept-drifting data stream to detect peer to peer botnet traffic. University of Texas at Dallas Technical Report# UTDCS-05- 08
29. Mehra P (2012) A brief study and comparison of snort and bro open source network intrusion detection systems. Int J Adv Res Comput Commun Eng 1(6):383–386
30. Nagaraja S (2014) Botyacc: unified p2p botnet detection using behavioural analysis and graph analysis. In: Computer security-ESORICS 2014. Springer, pp 439–456
31. Narang P, Hota C, Venkatakrishnan V (2014) Peershark: flow-clustering and conversation-generation for malicious peer-to-peer traffic identification. EURASIP J Inf Secur 2014(1):1–12
32. Narang P, Khurana V, Hota C (2014) Machine-learning approaches for p2p botnet detection using signal-processing techniques. In: Proceedings of the 8th ACM international conference on distributed event-based systems. ACM, pp 338–341
33. Narang P, Ray S, Hota C, Venkatakrishnan V (2014) Peershark: detecting peer-to-peer botnets by tracking conversations. In: Security and privacy workshops (SPW), 2014. IEEE, pp 108–115
34. Narang P, Reddy JM, Hota C (2013) Feature selection for detection of peer-to-peer botnet traffic. In: Proceedings of the 6th ACM India computing convention, pp 16:1–16:9
35. Puttaswamy KP, Zheng H, Zhao BY (2009) Securing structured overlays against identity attacks. IEEE Trans Parallel Distrib Syst 20(10):1487–1498
36. Rahbarinia B, Perdisci R, Lanzi A, Li K (2013) Peerrush: mining for unwanted p2p traffic. Detection of intrusions and malware, and vulnerability assessment. Springer, Berlin, pp 62–82
37. Ratnasamy S, Francis P, Handley M, Karp R, Shenker S (2001) A scalable content-addressable network, vol 31. ACM
38. Reddy JM, Hota C (2013) P2p traffic classification using ensemble learning. In: Proceedings of the 5th IBM collaborative academia research exchange workshop. ACM, p 14
39. Roesch M et al (1999) Snort: lightweight intrusion detection for networks. LISA 99:229–238
40. Rowstron A, Druschel P (2001) Pastry: scalable, decentralized object location, and routing for large-scale peer-to-peer systems. In: Middleware 2001. Springer, pp 329–350
41. Ruehrup S, Urbano P, Berger A, D'Alconzo A (2013) Botnet detection revisited: Theory and practice of finding malicious p2p networks via internet connection graphs. In: 2013 IEEE conference on computer communications workshops (INFOCOM WKSHPS). IEEE, pp 435–440
42. Schoof R, Koning R (2007) Detecting peer-to-peer botnets. University of Amsterdam. Technical report
43. Sen S, Spatscheck O, Wang D (2004) Accurate, scalable in-network identification of p2p traffic using application signatures. In: Proceedings of the 13th international conference on World Wide Web. ACM, pp 512–521
44. Singh A et al (2006) Eclipse attacks on overlay networks: threats and defenses. In: IEEE INFOCOM, Citeseer
45. Steiner M, En-Najjary T, Biersack EW (2007) Exploiting kad: possible uses and misuses. ACM SIGCOMM Comput Commun Rev 37(5):65–70
46. Stover S, Dittrich D, Hernandez J, Dietrich S (2007) Analysis of the storm and nugache trojans: P2p is here. USENIX 32(6):18–27
47. Tegeler F, Fu X, Vigna G, Kruegel C (2012) Botfinder: finding bots in network traffic without deep packet inspection. In: Proceedings of the 8th international conference on emerging networking experiments and technologies. ACM, pp 349–360

48. Trifa Z, Khemakhem M (2012) Taxonomy of structured p2p overlay networks security attacks. World Acad Sci, Eng Technol 6(4):460–466
49. Yu H, Kaminsky M, Gibbons PB, Flaxman A (2006) Sybilguard: defending against sybil attacks via social networks. ACM SIGCOMM Comput Commun Rev 36(4):267–278
50. Yue X, Qiu X, Ji Y, Zhang C (2009) P2p attack taxonomy and relationship analysis. In: 11th International conference on advanced communication technology, 2009. ICACT 2009, vol 2. IEEE, pp 1207–1210
51. Zhang R, Zhang J, Chen Y, Qin N, Liu B, Zhang Y (2011) Making eclipse attacks computationally infeasible in large-scale dhts. In: 2011 IEEE 30th International performance computing and communications conference (IPCCC). IEEE, pp 1–8

Application-Level Benchmarking of Big Data Systems

Chaitanya Baru and Tilmann Rabl

Abstract The increasing possibilities to collect vast amounts of data—whether in science, commerce, social networking, or government—have led to the "big data" phenomenon. The amount, rate, and variety of data that are assembled—for almost any application domain—are necessitating a reexamination of old technologies and development of new technologies to get value from the data, in a timely fashion. With increasing adoption and penetration of mobile technologies, and increasing ubiquitous use of sensors and small devices in the so-called *Internet of Things*, the big data phenomenon will only create more pressures on data collection and processing for transforming data into knowledge for discovery and action. A vibrant industry has been created around the big data phenomena, leading also to an energetic research agenda in this area. With the proliferation of big data hardware and software solutions in industry and research, there is a pressing need for benchmarks that can provide objective evaluations of alternative technologies and solution approaches to a given big data problem. This chapter gives an introduction to big data benchmarking and presents different proposals and standardization efforts.

1 Introduction

As described in [11], database system benchmarking has a rich history, from the work described in the paper entitled, *A Measure of Transaction Processing Power* [1], to the establishment of the Transaction Processing Council in 1988, and continuing into the present. The pressing need for database benchmark standards was recognized in the mid to late 1980s, when database technology was relatively new,

C. Baru (✉)
San Diego Supercomputer Center, University of California, San Diego, USA
e-mail: baru@sdsc.edu

T. Rabl
bankmark, Passau, Germany
e-mail: tilmann.rabl@bankmark.de

© Springer India 2016   189
S. Pyne et al. (eds.), *Big Data Analytics*, DOI 10.1007/978-81-322-3628-3_10

and a number of companies were competing directly in the database systems software marketplace. The initial efforts were simply on persuading competing organizations to utilize the *same* benchmark specification—such as the *DebitCredit* benchmark introduced in [12]. However, the benchmark results were published directly by each company, often eliminating key requirements specified in the original benchmark. Thus, there was need for a standard-based approach, along with a standard organization that could uniformly enforce benchmark rules while also certifying results produced, thereby providing a stamp of approval on the results. Thus was born the *Transaction Processing Performance Council, or TPC,* in 1988. With its early benchmarks, such as TPC-C, TPC-D, and TPC-H, the TPC was successful in producing widely used benchmark standards that have led directly to database system product improvements and made a real impact on product features.

With the rapid growth in big data[1] applications, and vendor claims of hardware and software solutions aimed at this market, there is once again a need for objective benchmarks for systems that support big data applications. In the emerging world of "big data," organizations were once again publishing private benchmark results, which claimed performance results that were not audited or verified. To address this need, a group from academia and industry (including the authors) organized the first *Workshop on Big Data Benchmarking (WBDB)* in May 2012, in San Jose, California. Discussions at the WBDB workshop covered the full range of issues, and reinforced the need for benchmark standards for big data. However, there was also recognition of the challenge in defining a commonly agreed upon set of big data application scenarios that could lead toward benchmark standards.

In retrospect, the early TPC benchmarks had an easier time in this regard. Their initial focus was transaction processing—typically defined by *insert, update, delete,* and *read* operations on records or fields within records. Examples are point-of-sale terminals in retail shopping applications, bank teller systems, or ticket reservation systems. Subsequent benchmarks extended to SQL query processing with relational database systems. Big data applications scenarios are, however, much more varied than transaction processing plus query processing. They may involve complex transformation of data, graph traversals, data mining, machine learning, sequence analysis, time series processing, and spatiotemporal analysis, *in addition to* query processing. The first challenge, therefore, is to identify application scenarios that capture the key aspects of big data applications. Application-level data benchmarks are *end-to-end benchmarks* that strive to cover the performance of all aspects of the application, from data ingestion to analysis.

A benchmark specification must pay attention to multiple aspects, including

(a) The so-called *system under test (SUT),* i.e., the system or components that are the focus of the testing, which may range from a single hardware or software component to a complete hardware or software systems.

---

[1]The term "big data" is often written with capitals, i.e. Big Data. In this paper, we have chosen to write this term without capitalization.

(b) The types of *workloads*, from application-level to specific component-level operations. *Component benchmarks* focus on specific operations; examples are I/O system benchmarks, graphics hardware benchmarks, or sorting benchmarks. The types of workloads range from very simple *micro-benchmarks*, which test a certain type of operation, to complex application simulations or replay of real workloads.

(c) The *benchmarking process*. *Kit-based benchmarks* provide an implementation or suite of tools that automates the benchmarking process. *Specification-based benchmarks* describe the detailed benchmarking process and allow for different implementations of the benchmark. The former are typically used in component benchmarks, while the latter are used for database, end-to-end benchmarks.

(d) The *target audience*. Benchmark details and especially the representation of the results may differ depending upon the target audience for the benchmark. End users and product marketing may require results that are easily comparable with realistic workloads. Performance engineers may prefer workloads that cover typical modes of operation. System testers may want to cover *all* modes of operation, but also need deep insights into the system behavior. Big data benchmarks exist in all of the forms described above.

While benchmarks can only *represent* real-world scenarios—and are not the real-world scenarios themselves—they nonetheless play an essential role. They can represent a broad class of application needs, requirements, and characteristics; provide repeatability of results; facilitate comparability among different systems; and provide efficient implementations. A good benchmark would represent the important aspects of real-world application scenarios as closely as possible, provide repeatability and comparability of results, and would be easy to execute.

The rest of this paper is structured as follows. Section 2 provides examples of some big data application scenarios. Section 3 describes useful benchmark abstractions that represent large classes of big data applications. Section 4 describes the approach taken by different benchmark standards, such as TPC and SPEC. Section 5 describes current benchmarking efforts, and Sect. 6 provides a conclusion.

2 Big Data Application Examples

While there is broad consensus on the potential for big data to provide new insights in scientific and business applications, characterizing the particular nature of big data and big data applications is a challenging task—due to the breadth of possible applications. In 2014, the US *National Institute for Standards and Technologies (NIST)* initiated a NIST Big Data Public Working Group (NBD-PWG) in order to tackle the issue of developing a common framework and terminology for big data and big data applications.[2] As documented in the NBD-PWG volume on *Use Cases and General Requirements*,[3] the range of real-world big data applications is broad, and includes collection and archiving of data; use in trading and financial sector; delivery of streaming content in a variety of applications scenarios including, for example, security, entertainment, and scientific applications; indexing to support web search; tracking data streams related to shipping and delivery of physical items (e.g., by FedEX, UPS, or US Postal Service); collection and analysis of data from sensors, in general; personalized health, precision medicine and other applications in healthcare and life sciences (including electronic medical records, pathology, bioimaging, genomics, epidemiology, people activity models, and biodiversity); deep learning with social media and a variety of other data; processing for driverless cars; language translation; smart grids; and others.

One of the example applications in the list of NIST Uses Cases is the *Genome in a Bottle Consortium*[4]—a public–private–academic consortium hosted by NIST to develop the technical infrastructure needed to enable translation of whole human genome sequencing to clinical practice. A current pilot application at NIST, employing open-source bioinformatics sequencing software on a 72-core cluster, has generated 40 TB of data. However, DNA sequencers will be able to generate ~300 GB compressed data per day in the near future. An individual lab will easily be able to produce petabytes of genomics data in future.

In another example, the Census Bureau is considering the use of advanced techniques to conduct demographic surveys, whose costs are increasing even as survey responses decline. These techniques include recommendation systems to help improve response rates; use of "mash ups" of data from multiple sources; and the use of historical survey administrative data, to help increase the overall survey data quality while simultaneously reducing their costs. As described in [ref], the US Census is planning to use big data in multiple ways including for "*Address canvassing*," i.e., to validate, correct, or delete existing Census Bureau *addresses*; to assist in the planning of the survey itself; and to assist in nonresponse follow-up.

A third use case deals with *large-scale deep learning* models. Deep learning algorithms employ architectures with multiple levels of nonlinear transformations, where the output of each layer is provided as input to the next. The demonstrated ability of deep learning models to extract representations and information from big data have made them attractive approaches for a number of big data applications including, for example, image and speech recognition. With increasingly sophisticated speech-based interfaces to computers and devices, and the need for image processing/recognition in applications like self-driving cars, the big data needs will only increase in this area.

While these examples provide a glimpse of the vast potential of data-intensive approaches in real applications, they also illustrate the challenges in defining the

---

[2]NIST Big Data Public Working Group, http://bigdatawg.nist.gov/home.php.
[3]NIST NBD-PWG Use Cases and Requirement, http://bigdatawg.nist.gov/usecases.php.
[4]*Genome in a Bottle Consortium*, https://sites.stanford.edu/abms/giab.

scope of the benchmarking problem for big data applications. The next section provides two specific approaches to tackling this issue.

3 Levels of Abstraction

Since the scope of big data applications can be vast—as described in the previous section—it is important to develop "abstractions" of real applications, in order to then develop benchmark specifications, which are based on those abstractions. The two abstractions described in this section are based on (1) extending the familiar *data warehouse* model to include certain big data characteristic in the data and the workload and (2) specifying a *pipeline* of processing, where data is transformed and processed in several steps. The specifics of each step may be different for different applications domains.

As described in Sect. 1, TPC benchmarks provide an application-level abstraction of business applications, such as transaction processing and/or complex query processing. TPC benchmarks, such as TPC-C, TPC-H, TPC-DS,[5] model retail transaction processing and data warehousing environments. The database model and workload provide a representative view of a specific business application scenario. Nonetheless, the results of such benchmarks can be used as a guide for a variety of other application (non-business) use cases with similar characteristics. Following this example, we present two models for big data benchmarks, in this section. The first, *BigBench,* follows the TPC model; while the second, *Deep Analytics Pipeline*, is a model based on a generalization of big data processing pipelines.

3.1 BigBench

BigBench is a big data analytics benchmark based on TPC-DS. Its development was initiated at the first *Workshop on Big Data Benchmarking* in May 2012 [6]. BigBench models a retail warehouse that has two sale channels: web sales and store sales. An excerpt of the data model is shown in Fig. 1. In order to appropriately reflect the big data use case, BigBench features not only structured data but also semi-structured and unstructured data. These data sections contain dependencies. For example, the web log, or clickstream data (semi-structured), has references to the SALES table. There are thirty queries specified in the BigBench workload, which cover a broad variety of analytics representing the different big data levers that were identified by Manyika et al. [8]. The queries use cases cover business aspects of marketing, merchandising, operations, supply chains, and reporting in a typical enterprise.

The BigBench processing model is targeted at batch analytics. The complete benchmark process consists of three stages: data generation, data load, a *Power*

---

[5]TPC, http://www.tpc.org/.

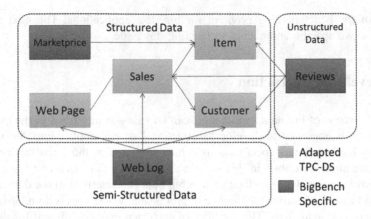

Fig. 1 BigBench data model

Test, and a *Throughput Test.* The data generation step generates the complete data set in different flat file formats. However, it is not part of the measured benchmark run. In the loading stage, data can be loaded into the system under test. The *Power Test* is performed by a serial execution of all thirty queries in the workload, while the *Throughput Test* consists of running a preselected number of serial streams, each of which is a permutation of the thirty queries. The total number of queries run is divided by the benchmark runtime to obtain a queries-per-hour metric.

BigBench has gained widespread attention in industry and academia and is currently in process to be standardized by the TPC. Several extensions were proposed [5].

3.2 Data Analytics Pipeline

Another proposal for a big data benchmark, referred to as the *Deep Analytics Pipeline,* is based on the observation that many big data applications are required to process streams of data, which requires a pipelined approach to data processing [4]. Figure 2, from [4], shows the steps in such a pipeline. This pipelined approach to processing big data is, indeed, a generic processing "pattern" that applies to a wide range of application scenarios. For example, processing of user clickstream data to predict ad placement in online advertising applications; predicting *customer churn* in, say, the banking industry based on user interactions as well as demographics, income, and other customer data; predicting insurance fraud based on a customer's activities; and predicting a patient's likelihood to visit the emergency room, based again on their activity and health information. All these examples involve using a variety of data related to the entity of interest, e.g., customer, patient, online user, etc., and detecting correlations between "interesting" outcomes and prior behavior.

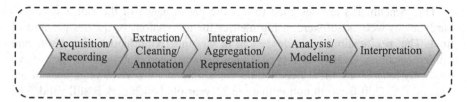

Fig. 2 Deep analytics pipeline

In essence, all such applications are performing a "user modeling" exercise based on transactional data and other historical and related information.

Pipelined data processing is also very typical of many, if not most, scientific applications as well. The processing pipeline in this case includes steps of data acquisition, data ingestion (data cleaning and normalization), data validation, and a variety of downstream analytics and visualization. The pipeline model is, thus, generic in nature. Different classes of application may be characterized by variations in the steps of the pipeline.

For benchmarking purposes, the pipeline stages are described based on their function, rather than in platform-specific terms. The pipeline stages could be implemented on a single hardware platform, or on distributed platforms.

4 Benchmarks Standards

Current industry standards provide two successful models for data-related benchmarks, TPC and SPEC, both of which were formed in 1988. TPC was launched with the objectives of creating standard benchmarks and a standard process for reviewing and monitoring those benchmarks [11]. SPEC was founded in the same year by a small number of workstation vendors who realized that the marketplace was in desperate need of realistic, standardized performance tests. The key realization was that an "ounce of honest data was worth more than a pound of marketing hype." Interestingly, the current community interest in big data benchmarking has the same motivation! Both organizations, TPC and SPEC, operate on a membership model. Industry as well as academic groups may join these organizations.

4.1 The TPC Model

TPC benchmarks are free for download; utilize standardized metrics for measuring transaction and query throughput; measure performance as well as price performance of given solutions; and have more recently introduced an energy metric, to measure performance versus energy consumption.

TPC benchmarks are designed to test the performance of the entire system—hardware as well as software, using metrics for transactions or queries per unit of time. The specifications are independent of the underlying hardware and software implementations. Over its history, TPC has demonstrated that its benchmarks have relatively long "shelf life." Benchmarks remain valid for several years and, in the case of TPC-C, it is worth noting that the 22-year-old benchmark is still valid. The benchmark measures transactions per minute for a scenario based on order-entry systems. The transactions include entering and delivering orders, recording payments, checking the status of orders, and monitoring the level of stock at warehouses. One of the keys to the longevity of the TPC-C benchmark is the rule for "data scaling," which is based on a "continuous scaling" model, where the number of warehouses in the database scales up with the number of transactions.

TPC benchmark "sponsors" may publish official TPC results for a fee. The publication rules require full disclosure of all information, including system configuration, pricing, and details of performance. Benchmark results are audited by an independent, third-party auditor.

TPC benchmarks that are query processing-oriented specify fixed database sizes at which the benchmark may be executed. These so-called database *scale factors* range from 1—representing a 1 GB "raw" database size—to 100,000 for a 100 TB raw database size. The most common scale factors for which benchmarks have been published are ranging from 100 to 10,000 (100 GB to 10 TB). Big data benchmarks may also need to adopt a similar scheme for database sizes.

4.2   The SPEC Model

SPEC benchmarks typically focus on specific functions or operations within a system, e.g., integer performance, sort performance. The benchmarks are typically server-centric, and test performance of small systems or components of systems. Unlike TPC, each benchmark defines its own metric, since different benchmarks may focus on different aspects of a system. As a result, they tend to have short shelf life—benchmarks can become obsolete with a new generation of hardware or system. The SPEC benchmark toolkits can be downloaded for a fee.

Publication of benchmark results by SPEC is free to members and subject to a modest fee for nonmembers. Unlike TPC, which incorporates third-party audits, SPEC benchmark results are peer reviewed. Also, unlike TPC, which requires extensive, full disclosure, SPEC requires only a disclosure summary—partly because the benchmark is at a component level within a system. The following table—from a tutorial on *Big Data Benchmarking* presented by Baru and Rabl at the IEEE Big Data Conference, 2014[6]—summarizes the features of TPC versus SPEC benchmarks.

---

[6]http://www.slideshare.net/tilmann_rabl/ieee2014-tutorialbarurabl?qid=d21f7949-d467-4824-adb1-2e394e9a2239&v=&b=&from_search=1.

TPC model	SPEC model
Specification based	Kit based
Performance, price, energy in one benchmark	Performance and energy in separate benchmarks
End-to-end benchmark	Server-centric benchmark
Multiple tests (ACID, load, etc.)	Single test
Independent review	Peer review
Full disclosure	Summary disclosure

4.3 Elasticity

An important requirement for big data systems is *elasticity*. For big data systems that deal with large data that are continuously growing, e.g., clicks streams and sensor streams, the system must be designed to automatically take advantage of additional resources, e.g., disks or nodes, as they are added to the system. Conversely, given the scale of many big data systems, there is a high probability of component failures during any reasonable workload run. The loss of a component in such a system should not lead to application failure, system shutdown, or other catastrophic results. The system should be "elastic" in how it also adopts to addition and/or loss of resources. Benchmarks designed for big data system should attempt to incorporate these features as part of the benchmark itself, since they occur as a matter of course in such systems. Currently, TPC benchmarks do require ACID tests (for testing atomicity, consistency, isolation, and durability properties of transaction processing and/or database systems). However, such tests are done outside the benchmark window, i.e., they are not a part of the benchmark run itself, but are performed separately.

5 Current Benchmarking Efforts

Several efforts have been created to foster the development of big data benchmarks. A recent paper summarizes a number of such efforts [7]. The first *Workshop on Big Data Benchmarking*, which was held in San Jose, California, USA, in May 2012, created one of the first community forums on this topic. In this workshop, 60 participants from industry and academia came together to discuss the development of industry standard big data benchmarks. The workshop resulted in two publications [2, 3] and the creation of the big data Benchmarking Community (BDBC), an open discussion group that met in biweekly conference calls and via online discussions. In mid-2014, with the creation of the SPEC Research Group on big data,[7] the BDBC group was merged with the SPEC activity. With the formation of the SPEC RG, weekly calls have been established, as part of the SPEC activity, with the weekly presentations alternating between open and internal calls. The open presentations cover new big data systems, benchmarking efforts, use cases, and related research. The internal calls are restricted to SPEC members only, and focus on discussion of big data benchmark standardization activities.

The WBDB series launched in 2012 has been continuing successfully, with workshops in the India (December 2012), China (July 2013), US (October 2013), Germany (October 2014), and Canada (June 2015). The next workshop, the 7th WBDB, will be held on December 14–15 in New Delhi, India. The WBDB workshops have, from the beginning, included participation by members of standards bodies, such as SPEC and TPC. Workshop discussions and papers presented at WBDB have led to the creation of the SPEC Research Group on big data, as mentioned earlier, and creation of the TPC Express Benchmark for Hadoop Systems (*aka* TPCx-HS), which is the first industry standard benchmark for Apache Hadoop compatible big data systems, and is based on the Terasort benchmark. The BigBench paper presented first at WBDB has also led to the formation of the TPC–BigBench subcommittee, which is working toward a big data benchmark based on a data warehouse-style workload.

Finally, an idea that has been discussed at WBDB is the notion of creating a BigData Top100 List[8], based on the well-known TOP500 list used for supercomputer systems. Similar to the TOP500, the BigData Top100 would rank the world's fastest big data systems—with an important caveat. Unlike the TOP500 list, the BigData Top100 List would include a price/performance metric.

6 Conclusion

Big data benchmarking has become an important topic of discussion, since the launching of the WBDB workshop series in May 2012. A variety of projects in academia as well as industry are working on this issue. The WBDB workshops have provided a forum for discussing the variety and complexity of big data benchmarking—including discussions of who should define the benchmarks, e.g., technology vendors versus technology users/customers; what new features should be include in such benchmarks, which have not been considered in previous performance benchmarks, e.g., *elasticity* and *fault tolerance*.

Even though this is a challenging topic, the strong community interest in developing standards in this area has resulted in the creation of the TPCx-HS benchmark; formation of the TPC-BigBench subcommittee; and the formation of

---

[7]SPEC RG big data - http://research.spec.org/working-groups/big-data-working-group.
[8]BigData Top100 - http://www.bigdatatop100.org/.

the SPEC research group on big data. Finally, a benchmark is only a formal, standardized representation of "typical" real-world workloads that allows for comparability among different systems. Eventually, users are interested in the performance of their specific workload(s) on a given system. If a given workload can be formally characterized, it could then be executed as a service across many different systems, to measure the performance of any system on that workload.

References

1. Anon et al (1985) A measure of transaction processing power. Datamation, 1 April 1985
2. Baru C, Bhandarkar M, Poess M, Nambiar R, Rabl T (2012) Setting the direction for big data benchmark standards. In: TPC-Technical conference, VLDB 2012, 26–28 July 2012, Istanbul, Turkey
3. Baru C, Bhandarkar M, Nambiar R, Poess M, Rabl T (2013) Benchmarking big data Systems and the BigData Top100 List. Big Data 1(1):60–64
4. Baru Chaitanya, Bhandarkar Milind, Nambiar Raghunath, Poess Meikel, Rabl Tilmann (2013) Benchmarking Big Data systems and the BigData Top100 List. Big Data 1(1):60–64
5. Baru C, Bhandarkar M, Curino C, Danisch M, Frank M, Gowda B, Jacobsen HA, Jie H, Kumar D, Nambiar R, Poess P, Raab F, Rabl T, Ravi N, Sachs K, Sen S, Yi L, Youn C (2014) Discussion of BigBench: a proposed industry standard performance benchmark for big data. In: TPC-Technical conference, VLDB
6. Ghazal A, Rabl T, Hu M, Raab F, Poess M, Crolotte A, Jacobsen HA (2013) BigBench: towards an industry standard benchmark for big data analytics. In: Proceedings of the 2013 ACM SIGMOD conference
7. Ivanov T, Rabl T, Poess M, Queralt A, Poelman J, Poggi N (2015) Big data benchmark compendium. In: TPC technical conference, VLDB 2015, Waikoloa, Hawaii, 31 Aug 2015
8. Manyika J, Chui M, Brown B, Bughin J, Dobbs R, Roxburgh C, Byers AH (2011) Big data: the next frontier for innovation, competition, and productivity. Technical report, McKinsey Global Institute. http://www.mckinsey.com/insights/mgi/research/technology_and_innovation/big_data_the_next_frontier_for_innovation
9. Rabl T, Poess M, Baru C, Jacobsen HA (2014) Specifying big data benchmarks. LNCS, vol 8163. Springer, Berlin
10. Rabl T, Nambiar R, Poess M, Bhandarkar M, Jacobsen HA, Baru C (2014) Advancing big data benchmarks. LNCS, vol 8585. Springer, Berlin
11. Shanley K (1998) History and overview of the TPC. http://www.tpc.org/information/about/history.asp
12. Serlin O. The History of DebitCredit and the TPC, http://research.microsoft.com/en-us/um/people/gray/benchmarkhandbook/chapter2.pdf

Managing Large-Scale Standardized Electronic Health Records

Shivani Batra and Shelly Sachdeva

Abstract Electronic health records (EHRs) contain data about a person's health history. Increasingly, EHRs have the characteristics of big data in terms of their volume, velocity, and variety (the 3 "V"s). Volume is a major concern for EHRs especially due to the presence of huge amount of null data, i.e., for storing sparse data that leads to storage wastage. Reducing storage wastage due to sparse values requires amendments to the storage mechanism that stores only non-null data, and also allows faster data retrieval and supports multidimensional heterogeneous data. Another area of concern regarding EHRs data is standardization. Standardization can aid in semantic interoperability that resolves the discrepancies in interpretation of health records among different medical organizations or persons involved. Various proposals have been made at the logical layer of relational database management system for managing large-scale standardized records in terms of data volume, velocity, and variety. Every proposed modification to logical layer has its pros and cons. In this chapter, we will discuss various aspects of the solutions proposed for managing standardized EHRs, and the approaches to adopt these standards. After efficient management of EHR data, analytics can be applied to minimize the overall cost of healthcare.

Keywords Electronic health records (EHRs) · Entity attribute value model · Storage schema · Access speed

S. Batra · S. Sachdeva (✉)
Department of Computer Science and Engineering, Jaypee Institute of Information
Technology University, Sector-128, Noida 201301, India
e-mail: shelly.sachdeva@jiit.ac.in

S. Batra
e-mail: ms.shivani.batra@gmail.com

© Springer India 2016
S. Pyne et al. (eds.), *Big Data Analytics*, DOI 10.1007/978-81-322-3628-3_11

1   Introduction

In today's world of digitization, a large amount of the generated data is stored electronically. Healthcare is a major domain with lots of data. For better utilization of health records, they are stored electronically. Not surprisingly, electronic health records (EHRs) data have a multitude of representations [1]. EHRs contain longitudinal (life-long) health-related information of a person. The contents may be structured, semi-structured, or unstructured, or a mixture of all these. In fact, these may be plain text, coded text, paragraphs, measured quantities (with values and units); date, time, date–time (and partial date/time); encapsulated data (multimedia, parsable content); basic types (such as, Boolean, state variable); container types (list, set); and uniform resource identifiers (URI). Scope of primary care is limited to one hospital, whereas an integrated care constitutes multiple hospitals. An integrated care EHR [2] is defined as "a repository of information regarding the health of a subject of care in computer processable form, stored and transmitted securely, and accessible by multiple authorized users. It has a commonly agreed logical information model which is independent of EHR systems. Its primary purpose is the support of continuing, efficient and quality integrated healthcare and it contains information which is retrospective (a historical view of health status and interventions), concurrent (a "now" view of health status and active interventions) and prospective (a future view of planned activities and interventions)."

Modern medicine is a data-intensive practice that generates large volumes of complex and heterogeneous data. Figure 1 shows the complexity of the healthcare domain. It is vast, consisting of patients, clinical concepts, relationships, terminology concepts, and terminology relationships. There are a large number of patients and clinical concepts. A clinical concept is basically a specific disease or fact such as, blood pressure or heart rate. One clinical concept might have some relationship (shown as EHR_relationship in Fig. 1) with other clinical concepts. For

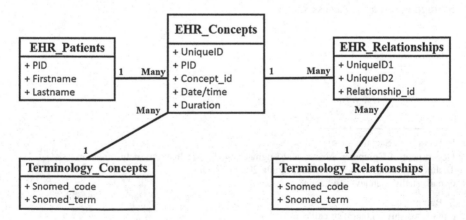

Fig. 1 Complexity of the Healthcare domain

example, increased heart rate associated with increased blood pressure may result in hypertension or cardiovascular disease.

The use of medical terminologies constitutes an important mechanism to standardize nomenclature and to introduce semantics to EHRs [1]. Clinicians often use strange terminology and a variety of codes to describe diagnosis. Also, physicians have different specialties. There are many standardized coding systems for medical terminology. Each defines its own set of terms and relation among terms. For example, the Systemized Nomenclature of Medicine–Clinical Terms (SNOMED CT) [3] coding system alone defines more than 400,000 clinical concepts, 1 million terms and 1.6 million relationships. A patient has one-to-many relation with clinical concepts. A clinical concept has one-to-many relation with EHR relationships. A terminology concept has one-to-many relation with clinical concepts. As healthcare data is becoming more complex, there is a need to handle such complexity. The challenges include capturing, storing, searching, sharing, and analyzing EHR data.

With every passing year, the healthcare costs keep rising. As per national health expenditure data maintained by the United States Centers of Medicare & Medicaid Services [4, 5], in 1963, US healthcare expenditure was just $35 billion that increased to $103 billion in 1973 with an average annual growth rate of 11.5 %. In year 2013, US health expenditure was approximately $3 trillion and it is estimated to be $5.1 trillion in year 2023, i.e., an average annual growth expected during the decade 2013–2023 is 5.8 %. One of the major reasons behind the increase in cost of healthcare is the exponential growth of EHRs data.

1.1 Electronic Health Records Growing BIG

While electronic medical records (EMRs) maintain information regarding a patient's medical records from a particular practice, EHRs capture longitudinal information related to an individual's overall health. EMRs are intended primarily for local use and hence lack in terms of providing semantic interoperability. Indeed, absence of support for semantic interoperability can hinder worldwide adoption of any healthcare information system.

EHRs offer a more general digital solution to the world of healthcare that otherwise runs on a chain of paper files. There has been a tremendous increase in the daily generation of EHR data. In an usual hospital scenario, a patient's vitals such as, temperature, heart beat, and blood pressure are recorded multiple times every day. Recording each and every detail of a patient leads to huge volume of medical data. With the exponential growth of EHRs, the proverbial 3 V's (Volume, Velocity, and Variety) of big data are becoming more relevant. Here, volume signifies the size of EHR data. Velocity refers to the access speed expected from a typical clinical information system, and variety to the diverse types of data structures involved in EHRs.

In 2012, worldwide digital healthcare data was estimated to be equal to 500 petabytes and is expected to reach 25,000 petabytes in 2020 [6]. It is expected that the size of health data will soon increase to zettabytes (equivalent 10^{21} gigabytes) and even yottabytes (10^{24} gigabytes) [7].

Considering the challenges of the healthcare domain, however, big data is not just about size. It also involves obtaining insights from complex, noisy, heterogeneous, longitudinal, and voluminous data. Goal of EHR is not just limited to capturing large amounts of data, but to improve patient care, patient heath, and to decrease overall cost. Trading, auditing, analyzing, and displaying of results are made possible by EHR data management and analytics.

Analyzing big data to obtain new insights is routinely conducted by pharmaceutical industry experts, providers, doctors, insurance experts, researchers, and customers. For instance, researchers can mine adverse health events data to detect the severity of effects upon taking certain combinations of drugs. Similarly, healthcare analysts and practitioners can efficiently handle and make inferences from voluminous and heterogeneous healthcare data to improve the health of patients while decreasing the costs of care. It has been estimated that big data analytics can help in saving approximately $300 billion per year in the US [8]. Analyzing EHRs big data can help in reducing waste and inefficiency in clinical operations, research and development, public health programs, evidence-based medicine, genomic medicine, pre-adjudication fraud analysis, device/remote monitoring, and predict health events and outcomes [9]. It has been reported that analysis of EHRs provided decision support and thus, improved healthcare services at approximately 350 hospitals which helped in saving estimated 160000 lives and reduced health spending by nearly $13.2 billion in the year 2008 [10].

For efficient analytics and cost reduction, the 3Vs of big data must be handled efficiently. For managing large volumes of EHR data constituting a variety of data types and demanding high access velocity, it is important that the storage issues be resolved first.

1.2 Storage Issues for EHRs

As EHR data volume increases, the need for efficiently managing the data is also becoming prominent. While dealing with EHR data, three issues—Standardization, Sparseness, and Volatility—are important from storage point of view.

Standardization: Standardization of EHRs is the key to support semantic interoperability. National Health Information Network (NHIN) of USA defines semantic interoperability as "the ability to interpret, and, therefore, to make effective use of the information so exchanged" [11]. Similarly, Institute of Electrical and Electronics Engineers (IEEE) Standard 1073 defines semantic interoperability as shared data types, shared terminologies, and shared codings [12]. Standards provide a definitional basis for interoperability [13]. Achieving standardization thus helps in global acceptability, interoperability, and higher usability [14].

Sparseness: One of the major concerns regarding the volume of EHRs is the presence of large number of null values that leads to sparseness. A typical schema for EHRs must constitute each and every attribute contributing to healthcare. However, whenever a patient visits a doctor, typically only a few of his/her vitals are recorded while the remaining EHR attributes are kept null (following an ideal traditional relational model approach). This presence of large number of null values leads to wastage of storage space.

Volatility: Healthcare is an evolving domain. Naturally, many new clinical concepts come into existence every year. With this ever-changing environment, there is a need for adopting a system that is flexible enough to accommodate the changes.

2 Storing EHRs: Achieving Standardization

Different healthcare organizations adopt different modes for storing their health records. Some organizations store their records on paper, while increasingly many are doing so digitally. With the shift to digital medium, EHRs now face the problem of standardization. Healthcare is a diverse domain in terms of terminologies and ontologies. Standards attempt to define common terminology and ontology for all with no ambiguity. Following a standard can provide a common unambiguous picture of the data to everyone irrespective of their location and language. Hence, EHRs must be standardized, incorporating semantic interoperability. They must be designed to capture relevant clinical data using standardized data definitions and standardized quality measures.

2.1 Existing Standards

Several standards offer guidelines which can be used to build a common format for EHR data. Some examples are openEHR Foundation [15]; Consolidated Health Informatics Initiative (CHI) [16]; Certification Commission for Healthcare Information Technology (CCHIT) [16]; Healthcare Information and Management Systems Society (HIMSS) [16]; International Organization for Standardization (ISO), American National Standards Institute (ANSI) [16]; Canada Health Infoway [16]; Health Level Seven International (HL7) [17]; ISO 13606 [18, 19]; and European Committee for standardization (CEN's TC 251) [20].

Standardized terminologies are required for healthcare applications to ensure accurate diagnosis and treatment. For example, SNOMED CT is one of the developing standards.

Recently, Health and Human Services (HSS) of United States favored adoption of standards for easy consumer access and information sharing without information blocking [21].

2.2 Dual-Model Approach

Dual-model [22] approach is proposed with the aim of separating information domain from knowledge domain. Dual-model approach divides the architecture of EHRs into two layers, namely, the reference model (RM) [22] and the archetype model (AM) [22]. The presence of the two layers provides flexibility of adding new information to the existing EHR system without affecting the existing clinical information system. Figure 2 shows the system implementing the dual-model approach. This approach segregates the responsibilities of an information technology (IT) expert from those of a domain expert. The IT expert addresses various data management and interoperability issues in RM. The clinical expert does not participate in software development process and provides clinical domain knowledge which is used by system during runtime for end users (such as, patients, doctors, nurses, and researchers). An analogy to help us understand the dual-model approach is Lego bricks. Lego bricks are toy plastic bricks (analogy RM) with studs which can be connected to other plastic bricks and used to construct various desired structures such as, buildings and vehicles (analogy AM). Similarly, in dual-model approach, RM defines the fixed part of EHRs system such as, data structures being adopted in clinical information system. AM adopts RM for defining the constraints related to the underlying medical concepts such as, the name of an attribute, its range, its unit, and the data structure that it follows. For example, the clinical concept blood pressure (BP) constitutes five attributes (systolic, diastolic, mean arterial, pulse pressure, and comments) among which four record various types of pressure and one records comments. Various types of pressures recorded follow the 'QUANTITY' data type (for quantifiable or measurable data), while the comment attribute follows 'TEXT' data type (for textual data). Semantics of 'QUANTITY' and 'TEXT' are well defined in RM by the IT experts.

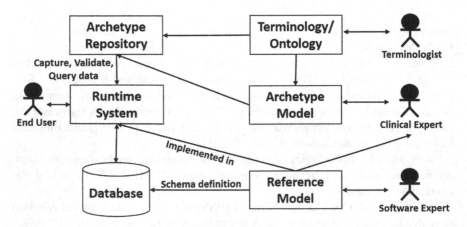

Fig. 2 System implementing dual-model approach

The clinical expert will only adopt the information defined in RM to build the medical concept based on their healthcare domain knowledge. Any modification in the future can be easily incorporated in AM without modifying RM. For example, a new type of pressure of data type 'QUANTITY' can be included without modifying the existing RM. AM delivers the medical concepts in the form of archetypes which are maintained in an archetype repository (as shown in Fig. 2). One of the online international standardized repositories is Clinical Knowledge Manager (CKM) [23] where archetypes based on the openEHR standard [15] are freely available for downloading. An archetype may logically include other archetypes, and/or may be a specialization of another archetype. Thus, they are flexible and vary in form. In terms of scope, they are general-purpose and reusable [1]. The clinical expert does not need to gather technical details about the classes defined in RM. User composability is provided by tools such as, archetype editor [24]. This enables the user, say, the clinical expert, to build archetype using the drag and drop functionality.

The ultimate goal of the dual-model approach is to bridge the IT expert and the healthcare domain specialist communities to foster interdisciplinary partnership. EHRs system based on the dual-model approach have the capability to incorporate new standardized data elements in a timely manner. With the dual-model approach, a small schema is defined in RM that can allow any type of domain knowledge in AM. As the domain knowledge expands, the archetype can accommodate it easily by versioning or creating new archetype using Archetype Definition Language (ADL) [25] following the same small schema (defined in RM). Once the archetypes are defined, they can be used for adopting the required standards. Thereafter, a data model must be selected to handle the issues of sparseness and volatility.

3 Managing EHRs: Dealing with Sparseness and Volatility

Considering the storage of EHRs, the most adopted data model is the traditional relational approach. Following the relational model, it requires each and every attribute (from different departments including administration, nursing, clinical, laboratory, radiology, and pharmacy) of a clinical information system to be stored. In real scenario, a patient episode at a medical facility will actually need a very small portion of the total number of attributes to be recorded. One reason for this may be that the patient may be diagnosed without any laboratory test, and thus, there is no data related to lab test would be recorded. When only small portions of attributes are recorded, the remaining unrecorded attributes are kept as null. This leads to high sparseness in EHR data. A relational model is not capable of dealing with sparseness efficiently.

One more problem with the relational model is volatility, i.e., the capability to add new attributes in an existing database schema. Addition of new attributes in a relational model is possible but upper limits on the total number of attributes are defined by the RDBMS (for example, maximum number of columns in DB2 and Oracle is 1012 [26]).

To deal with sparseness and volatility, data models (EAV, dynamic tables, OEAV, and OCOM) other than the relational model need to be adopted as the storing standard.

3.1 Entity Attribute Value Model

The Entity Attribute Value (EAV) model [27] stores data in the form of triplets comprising Entity, Attribute, and Value. Each data entry of relational model is stored as one row in EAV with the 'Entity' column storing the primary attribute. The 'Attribute' column stores the attribute name for which the data is being stored, and the 'Value' column stores the data value. The EAV model stores only non-null values of relational model.

To gain knowledge about the null values, EAV is supported with a metadata table (stored separately in the database) which stores information regarding the schema of the corresponding relational model. Figure 3a represents a typical relational model and Fig. 3b shows the equivalent EAV model. As shown in Fig. 3a, b, EAV model creates one row for each column value. Row 1 of the example in Fig. 3a is split into three rows in Fig. 3b each containing the value of one non-null attribute.

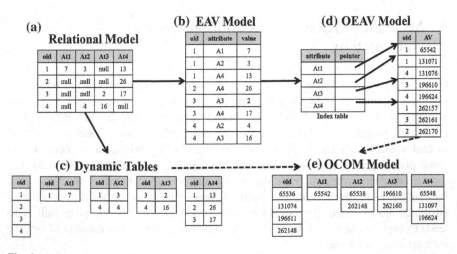

Fig. 3 Various data models

Advantages of EAV model: EAV model was proposed as an alternative for relational model to provide the following functionalities:

No Sparsity: EAV model reserves space only for non-null values. When information regarding null values is required, it can be accessed using metadata.

Volatility: It refers to the flexibility of adding new information to the existing knowledge domain. This flexibility is achievable in EAV by simply adding the new attributes name in the 'Attribute' column of EAV model.

Disadvantages of EAV model: A rigorous survey by the authors [28] found that the EAV model lacks certain aspects of the relational model as described below:

End user understanding: As depicted in Fig. 3b, gathering information from EAV model is complex when compared to a relational model. In relational model all information regarding an entity is stored in one row, whereas in EAV model it is scattered over multiple rows.

Query tools' unavailability: Till date, no query tools are available for managing EAV-based data. To compensate, existing RDBMSs are used for building, querying, and managing EAV-based data.

Handling multiple data types: The presence of a single 'Value' column in EAV model restricts the entry to one type of data. To store multiple data types, either multiple 'Value' columns are required, or multiple EAV tables (each storing a different data type) are required.

Retrieving data: Searching data related to an attribute or to an entity requires an exhaustive search of the whole EAV database. Thus, time taken to retrieve information from EAV database is much higher.

Size expansion with increase in non-null values: If a relational model has k non-null attributes for an entity, then the corresponding EAV model will contain k rows (one for each attribute) containing three entries each (Entity, attribute, and value). Hence, EAV-based storage will consume more space with increase in non-null entries.

3.2 Dynamic Tables

Dynamic table is an extension of decomposition model proposed by Copeland and Khoshafian [29, 30]. It works by dividing the n-ary relational model (i.e., containing n attributes) into n tables (as shown in Fig. 3c). Among these n tables, $n - 1$ tables are composed of two columns; first for the primary key ("oid") and second for a particular attribute other than the primary key. In addition to these $n - 1$ tables, there is one more table which contains only one column dedicated to store the contents of the primary key as in a relational model. The sole purpose of this additional table is to identify the existence of a particular entity in the database.

Advantages of Dynamic Tables: Dynamic tables overcome the problem of sparseness and volatility like the EAV model. In addition, dynamic tables are, in fact, superior to EAV because of the following advantages:

Retrieving attribute data: Searching information regarding an attribute requires access to the corresponding attribute table only. Time complexity of an attribute centric query in EAV model is $O(1)$.

Handling multiple data types: Different tables formed due to partitioning during the creation of dynamic tables store different attributes, and are thus, capable of handling different data types.

Size expansion with increase in non-null values: Dynamic tables also expand with the expansion of non-null values, but this expansion is less as compared to EAV.

End user understanding: Understanding dynamic tables is easier than EAV because they store one attribute in one table with the primary key associated with it.

Disadvantages of Dynamic Tables: The authors identified some limitations in dynamic tables which are described as follows [28]:

Retrieving a particular entity's data: It can be easily identified if a particular entity data is being stored in dynamic table or not by just searching for its id in oid table. However, to search the data corresponding to an existing entity (present in oid table) in dynamic tables is more complex since all tables (partitioned for dynamic tables) must be searched.

Memory consumption: Dynamic tables maintain only two entries per attribute value (the primary key and the attribute value), whereas EAV maintains three entries per attribute value (the primary key, the attribute name, and the attribute value). Hence, it may seem that storage requirements for dynamic tables are less than that of EAV, but that is not true. This fact was proven experimentally by the authors in [28]. More memory consumption in dynamic tables was due to the storage overhead of various tables (in dynamic tables) in comparison to just one table (in EAV).

3.3 Optimized Entity Attribute Value Model

Paul et al. [31] proposed the Optimized Entity Attribute Value (OEAV) model to improve search efficiency that was lacking in the EAV model. Constructing a database based on OEAV model first requires the data to be in EAV format. Next, attribute dictionary is constructed manually which defines numeric coding scheme for the attribute names. The combinations of numeric attributes codes and the corresponding values are stored as one single value under one column termed as, Attribute Value (AV) using the following steps:

1. Convert numeric attribute code and value to the equivalent binary representation comprising 16 bit each.
2. Concatenate both 16 bit codes to form a 32 bit code.
3. Now calculate the decimal equivalent of the 32 bit code.

For example, we want to calculate the AV value of (1, 11). According to step 1, 16 bit codes of (1, 11) are calculated, which is equivalent to (0000000000000001, 0000000000001011). These two 16 bit codes are combined to form a 32 bit code which is 00000000000000010000000000001011 (step 2) and its decimal equivalent is 65547 (step 3). The AV value, 65547 is now used as a replacement of (1, 11).

Finally, the transformed table is sorted based on the AV column and then the primary key (oid column in Fig. 3d). An index table is further created to point the first record (in the sorted table) of the corresponding attribute (in the index table). All the records corresponding to a particular attribute are found in the neighboring records; the 16 most significant digits of numeric codes will have the same value for same attribute. So, after sorting, all neighboring elements will have the same most significant digits until the records for the next attribute are started (having different most significant digits).

Advantages of OEAV model: In addition to the functionalities (of no sparseness or volatility) offered by EAV model, OEAV model achieves the following:

Retrieval of attribute data: Search efficiency of OEAV model is due to the presence of the index table. The index table points to the starting address of an attribute in the sorted table which makes OEAV a search efficient model.

Size expansion with increase in non-null values: The growth is less in OEAV model as compared to EAV model because in OEAV model only two entries (entity, AV) are created for each row, whereas in EAV model it is three (entity, attribute, and value).

Disadvantages of OEAV model: In addition to the limitations (of end user understandability, query tools unavailability, and handling multiple data types) in EAV model, the authors identified that OEAV model has the following limitations [28]:

Calculation overhead: Storage and retrieval of any information in OEAV require coding and decoding of AV column, respectively.

Insertion, modification, and deletion of records: Insertion of a new record is complex in the case of OEAV model, since records are finally sorted based on the value of the AV column. For inserting a new record, the database should be sorted again after the insertion. Doing so requires shifting of all records.

Retrieving an entity data: To retrieve the information of an entity, an exhaustive search of the whole database is required. This process is the same as it is in the EAV model. So, there is no significant improvement in searching data related to particular entity.

Handling multiple data types: The need for numeric coding for AV column requires numeric data. Hence, OEAV model is capable of storing only numeric data.

Dependency on machine: As shown in Fig. 3d, OEAV model constructs an index table which contains the pointer (address) to the first record of the corresponding attribute in the sorted table. If the database is reallocated to another machine, new address will be allocated to the OEAV database in the new machine. Thus, all pointer values in the index table will need to be reassigned to new address values, which makes the database machine dependent.

3.4 Optimized Column Oriented Model

Optimized Column Oriented Model (OCOM) [32] combines the features of dynamic tables and OEAV model (shown using dotted arrows in Fig. 3e). OCOM starts by dividing the relational table into multiple tables (each containing two columns, i.e., position number and attribute value) as in dynamic tables. After the division, each divided table is coded (two columns are coded in one column) using the coding mechanism of OEAV model.

Advantages of OCOM model: Apart from no sparseness and volatility, OCOM model shows advantages of both OEAV model and dynamic tables which are as follows [28]:

Memory consumption: OCOM model stores coded values of each entry (using position number and attribute value) of a attribute in a separate table. Hence, the total space consumed by OCOM is minimum among all the other compared models.

Size expansion with increase in non-null values: OCOM reserves exactly the same number of entries in a coded format as there were non-null entries. Thus, there is no growth in size with increase in non-null values rather the expansion is equivalent to the growth in non-null entries.

Dependency on the machine: Absence of index table (as in the OEAV model) makes OCOM model a machine-independent data model.

Disadvantages of OCOM model: OCOM model shows some limitations which are as follows:

Calculation overhead: Calculation overhead of OEAV model is also involved in OCOM model.

Retrieving data: Retrieval of data in OCOM model is quite complex due to the presence of coded values which need to be decoded before data retrieval.

Handling multiple data types: Similar to OEAV, OCOM also stores only numeric data as coding mechanism of OEAV model needs to be applied in OCOM model.

End user understanding: User cannot easily understand the coded format of OCOM model.

4 Modeling EHRs: A Big Data Perspective

The current section considers the big data issues in the modeling of large-scale EHR data, especially when it is not stored using relational model approach. EAV, dynamic tables, OEAV, and OCOM model are built with a primary aim of handling sparseness and volatility. The costs of EAV, dynamic tables, OEAV, and OCOM model are different in terms of data volume, velocity, variety, complex querying, complex data relationships, and view level support, as shown in Table 1.

Table 1 Comparison of data models from big data perspective

Big data requirements	Data models			
	EAV	Dynamic tables	OEAV	OCOM
Volume	Handle sparseness	Handle sparseness	Handle sparseness	Handle sparseness
Velocity	Slow data retrieval	Slow entity data retrieval	Complex data retrieval	Complex data retrieval
Variety	Requires multiple data type tables	Stores multiple types of data	Stores only numeric data	Stores only numeric data
Complex query	Requires modification in SQL structure	Requires little modification in SQL structure	Highly complex	Highly complex
Complex data relationship	Maintained using EAV/CR approach	Maintained implicitly	Not handled currently	Not handled currently
View level support	Require PIVOT operation	Require JOIN operation	Require PIVOT operation after decoding	Require JOIN operation after decoding

EAV, dynamic tables, OEAV, and OCOM model are capable of handling sparseness in data. They can make space for storing more data, and thus, handling EHRs' increase in volume.

One of the most prominent challenges for adopting EAV, dynamic tables, OEAV, and OCOM model is the unavailability of query tools. Adopting existing RDBMS for these models requires modifications at the logical level, which makes data accessing complex.

Managing the velocity of accessing data becomes complex in case of OEAV and OCOM due to the requirement of decoding data, whereas it is simpler in case of EAV than OEAV and OCOM, but is still slow. The most efficient model in terms of velocity among the discussed models is dynamic tables, which can access attribute data very fast, yet is slow for entity-related data.

Designing of EAV schema supports only one type of data that hinders the variety aspect of EHRs. OEAV and OCOM are designed only for numeric data which also have same drawback. Only dynamic tables satisfy the variety aspect of EHRs. The most important aspects that need to be handled by EAV, dynamic tables, OEAV, and OCOM models are described below.

4.1 Complex Query Requirement

Query must provide the most efficient means to gain access to a database. It allows for coding as needed by a user to gain access to a portion of database in a specified condition. Queries in healthcare domain are not only restricted to patient-related information (i.e., clinical queries) but may also extend to those related to population (i.e., epidemiological queries). The main reason for using relational model over other data models is the absence of query support. Many relational database management systems (RDBMSs) available in the market support relational models but none exists for EAV, dynamic tables, OEAV, and OCOM. To use a RDBMS for querying a database stored as EAV, dynamic tables, OEAV, and OCOM, the SQL query structure needs to be modified. For example, to access a particular attribute in relational model and the equivalent EAV model, the following queries will be used.

In Relational Model
select < attribute_name > from < table_name >;
In EAV Model
select value from < EAV_table_name > where attribute = "<attribute_name>";

4.2 Complex Data Relationship

Relationships among data in RDBMS is maintained using referential constraint, i.e., by combination of primary key and foreign key. In case of EAV, the primary key concept makes it difficult due to replication of the entity identifier in the 'Entity' column. To maintain relationships in EAV model, an extension to EAV called entity–attribute–value/class relationship (EAV/CR) [33] was proposed. In EAV/CR model, relationships among the classes are maintained using either the concept of primary and foreign keys or through the use of a special table that is used only for maintaining relationships. The main weakness of the EAV/CR model is the requirement for high data storage which can lead to problems such as, memory shortage and network congestion when applied on worldwide level to achieve semantic interoperability. A similar approach can be proposed for OEAV and OCOM. Dynamic tables support referential integrity constraint (foreign key) of the relational approach. So, there is no need of making any amendments to dynamic tables.

4.3 View Level Requirements

Users of databases are comfortable with using relational model where data related to a particular entity appears in one row and data related to one attribute in one column. With relational model, the user can have direct and simple interpretation of data. This is unlike the scenario of a database stored as EAV, dynamic tables, OEAV, or OCOM. For instance, to access data related to particular entity or attribute, the whole EAV database needs to be accessed.

To address the issue of complexity at view level of the RDBMS for EAV data, it should be converted into the relational model. The conversion of EAV into relational model can be achieved thorough pivoting [34]. However, pivoting involves the JOIN operation which has to be performed on different views of EAV database. Since join is a costly operation, it also makes pivoting very costly.

To make dynamic tables more understandable, all divisions (multiple tables) need to be joined back. Similar processes of pivoting and joining has to be applied to OEAV and OCOM after decoding the values stored therein.

The process to enable end user understanding could be of concern when the whole database needs to be accessed. In realistic scenario, the end user is authorized only to view a very small portion of the whole database. Hence, pivoting and joining can be applied accordingly, on a very small portion of the database, to ensure the positive aspects of the adopted model (EAV, dynamic tables, OEAV, or OCOM) while allowing significant end user understanding.

5 Applying Analytics on EHRs

The big data model provides a framework for managing data resources, and specifies the underlying architecture which can help in extracting knowledge from EHRs data at a reasonable computing expense. The architecture presented in Fig. 4 provides a visual mechanism to manage standardized EHRs for optimizing memory use (Volume), faster access speed (Velocity), and storing heterogeneous data (Variety). In particular, the management of standardized EHRs from a big data perspective is divided into four layers (Fig. 4).

EHRs constitute wide variety of data including structured data, unstructured clinical notes, clinical images, genetic data, epidemiology, and behavioral data originated from various departments such as, nursing, imaging lab, pathology labs, and the local hospital, as shown in Layer 1 of Fig. 4. EHRs collected from different sources are integrated for better understanding of the clinical situation. The large-scale EHR data follow different storage formats that can hinder semantic interoperability. To achieve semantic interoperability, standardization is essential.

Layer 2 of the architecture presents the way standards can be adopted for EHRs using dual-model approach. Reference model (RM) of the dual-model approach specified the stable part of clinical information system designed by IT experts. Based on RM, the dynamic part of clinical information system is constructed called archetype model (AM). Here, dual-model approach is described using the analogy of lego blocks. Lego blocks are produced as basic shape blocks (as in RM) using which any complex shape (as in AM) can be build. Any modification to the complex shapes built can be achieved with the existing basic blocks.

EHRs are highly sparse, which drives the choice for the suitable storage model. For handling data volume efficiently for highly sparse and volatile EHRs, relational model is not well suited. Layer #3 of Fig. 4 presents storage options other than a relational model, such as, EAV, dynamic tables, OEAV, and OCOM, which are designed with the aim of handling sparseness and adopting any change to the existing schema more flexibly.

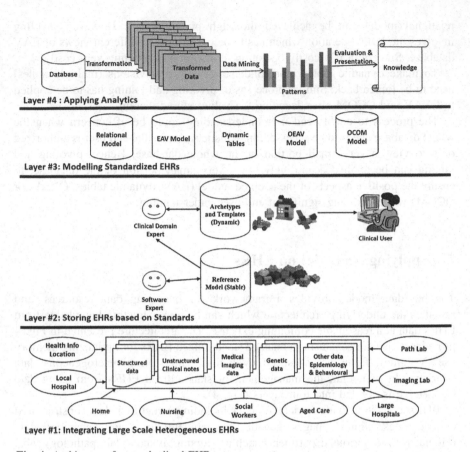

Fig. 4 Architecture for standardized EHRs management

Finally, we must have the required analytics applications that preserve these models, so that the models are able to display and store the latest data.

Data mining (shown in Layer #4) is a very useful process to discover patterns that lie hidden otherwise. Many tools such as, WEKA [35] are available for mining. Mining data in relational model is easy as the tools available in market are readily applicable. Currently, no major tools are available for mining EAV, dynamic tables, OEAV, and OCOM modeled data.

To mine databases using EAV, dynamic tables, OEAV, and OCOM models, the data should be converted into the relational format after which standard mining tools can be used easily for data analytics. Converting a whole database back into relational remains a matter of concern. In a realistic scenario, only a portion of data satisfying a particular condition may be required for mining. For example, identification of the relationship of high blood pressure patient with diabetes problem may require data only corresponding to the high blood pressure patients. So, the cost of pivoting may be compensated as only a small portion of data is converted back into relational model for mining. A detailed architecture for preparing standardized data for the purpose of mining is presented in [36].

6 Summary and Conclusion

Data is fastly becoming very large for many domains. Healthcare data is big not only because of its volume, but also due to its complex, diverse, and temporal nature. EHRs as big data require that the 3Vs must be handled carefully along with the issues of standardization, high sparsity, and volatility. This chapter explored various issues in the EHR domain. In doing so, it was found that the traditional relational model is not a suitable option for storing standardized EHRs.

Data models such as, EAV, dynamic tables, OEAV, and OCOM try to overcome the shortcomings of relational model with respect to the EHR requirements. These data models provide functionality of no sparseness and volatility for standardized EHRs, but they still lack in various aspects, among which the most common is entity centric query, query tool unavailability, and end user understandability. To improve end user understandability, pivoting and joining mechanisms need to be adopted to bring a small and selective data portion (as acquired data by the end user) to a better interpreted format (the relational model). Pivoting/joining is also required when mining and analytics are to be performed on EAV, dynamic tables, OEAV, and OCOM modeled databases. Analytics performed on EHRs can be useful for various users (such as, pharmaceutical industry experts, providers, doctors, insurance experts, researchers, and customers). It may lead to more efficient decision support while reducing overall decision-making errors and healthcare cost.

References

1. Sachdeva S, Bhalla S (2012) Semantic interoperability in standardized electronic health record databases. ACM J Data Inf Qual 3(1):Article 1
2. ISO/TC 215 Technical Report (2003) Electronic health record definition, scope, and context, 2nd draft, Aug
3. SNOMED. Clinical terms.Systematized nomenclature of medicine. http://www.snomed.org/documents/-snomedoverview.pdf
4. Hartman M, Martin AB, Lassman D, Catlin A (2015) National health expenditure accounts team. National health spending in 2013: growth slows, remains in step with the overall economy. Health Affairs (Millwood) 34(1):150–160
5. Keehan SP, Cuckler GA, Sisko AM, Madison AJ, Smith SD, Stone DA, Poisal JA, Wolfe CJ, Lizonitz JM (2015) National health expenditure projections, 2014–24: spending growth faster than recent trends. Health Affairs (Millwood) 34:1407–1417
6. Hersh W, Jacko JA, Greenes R, Tan J, Janies D, Embi PJ, Payne PR (2011) Health-care hit or miss? Nature 470(7334):327–329

7. IHTT: transforming health care through big data strategies for leveraging big data in the health care industry (2013). http://ihealthtran.com/wordpress/2013/03/iht%C2%B2-releases-big-data-research-reportdownload-today/
8. IBM: large gene interaction analytics at University at Buffalo, SUNY (2012) http://public.dhe.ibm.com/common/ssi/ecm/en/imc14675usen/IMC14675USEN.PDF
9. Raghupathi W, Raghupathi V (2014) Big data analytics in healthcare: promise and potential. Health Inf Sci Syst 2(1):3–10
10. Premier QUEST: https://www.premierinc.com/premier-quest-hospitals-outperform-national-mortality-averages-10-study-shows/. Accessed 03 2016
11. NHIN: Interoperability for the National Health Information Network (2005) IEEE-USA Medical Technology Policy Committee/Interoperability Working Group E-books
12. Kennelly RJ (1998) IEEE 1073, Standard for medical device communications. In: Proceedings of the IEEE systems readiness technology conference (Autotestcon '98), IEEE, Los Alamitos, CA, pp 335–336
13. Atalag K, Kingsford D, Paton C, Warren J (2010) Putting health record interoperability standards to work. J Health Inf 5:1
14. Batra S, Parashar HJ, Sachdeva S, Mehndiratta P (2013) Applying data mining techniques to standardized electronic health records for decision support. In: Sixth international conference on contemporary computing (IC3), pp 510–515. doi:10.1109/IC3.2013.6612249
15. OpenEHR Community. http://www.openehr.org/. Accessed 04 2015
16. EHR Standards. Electronic health records standards. http://en.wikipedia.org/wiki/Electronic health record Standards. Accessed 04 2015
17. HL7. Health level 7. www.hl7.org. Accessed 04 2015
18. ISO 13606-1. 2008. Health informatics: Electronic health record communication. Part 1: RM (1st Ed.). http://www.iso.org/iso/home/store/catalogue_tc/catalogue_detail.htm?csnumber=40784
19. ISO 13606-2. 2008. Health informatics: Electronic health record communication. Part 2: Archetype interchange specification (1st Ed.). http://www.iso.org/iso/home/store/catalogue_tc/catalogue_detail.htm?csnumber=62305
20. CEN TC/251. European standardization of health informatics. ENV 13606 Electronic Health Record Communication. http://www.centc251.org/
21. HHS.gov: http://www.hhs.gov/about/news/2016/02/29/hhs-announces-major-commitments-healthcare-industry-make-electronic-health-records-work-better.html. Accessed 03 2016
22. Beale T, Heard S (2008) The openEHR architecture: Architecture overview. In the openEHR release 1.0.2, openEHR Foundation
23. CKM. Clinical Knowledge Manager. http://www.openehr.org/knowledge/. Accessed 12 2014
24. Ocean Informatics—Knowledge Management. http://oceaninformatics.com/solutions/knowledge_management. Accessed 01 2016)
25. Beale T, Heard S (2008) The openEHR archetype model—archetype definition language ADL 1.4. In the openEHR release 1.0.2, openEHR Foundation
26. Xu Y, Agrawal R, Somani A (2001) Storage and querying of e-commerce data. In: Proceedings of the 27th VLDB conference, Roma, Italy
27. Dinu V, Nadkarni P (2007) Guidelines for the effective use of entity-attribute-value modeling for biomedical databases. Int J Med Inform 76:769–779
28. Batra S, Sachdeva S (2014) Suitability of data models for electronic health records database, big data analytics
29. Copeland GP, Khoshafian SN (1985) A decomposition storage model. In: Proceedings of the 1985 ACM SIGMOD international conference on management of data, pp 268–279
30. Khoshafian S, Copeland G, Jagodis T, Boral H, Valduriez P (1987) A query processing strategy for the decomposed storage model. In: Proceedings of the Third international conference on data engineering, pp 636–643
31. Paul R et al (2009) Optimized entity attribute value model: a search efficient representation of high dimensional and sparse data. IBC 3(9):1–6. doi:10.4051/ibc.2011.3.3.0009

32. Paul R, Hoque ASML (2010) Optimized column-oriented model: a storage and search efficient representation of medical data. Lecture Notes Computer Science, vol 6266, pp 118–127
33. El-Sappagh SH et al (2012) Electronic health record data model optimized for knowledge discovery. Int J Comput Sci 9(5, 1):329–338
34. Dinu V, Nadkarni P, Brandt C (2006) Pivoting approaches for bulk extraction of Entity–Attribute–Value data. Comput Methods Programs Biomed 82:38–43
35. Garner SR (1995) WEKA: The Waikato environment for knowledge analysis. In: Proceedings of the New Zealand computer science research students conference. The University of Waikato, Hamilton, New Zealand, pp 57–64
36. Batra S, Sachdeva S, Mehndiratta P, Parashar HJ (2014) Mining standardized semantic interoperable electronic healthcare records. In: First international conference, ACBIT 2013, Aizu-Wakamatsu, Japan, CCIS 404, pp 179–193

Microbiome Data Mining for Microbial Interactions and Relationships

Xingpeng Jiang and Xiaohua Hu

Abstract The study of how microbial species coexist and interact in a host-associated environment or a natural environment is crucial to advance basic microbiology science and the understanding of human health and diseases. Researchers have started to infer common interspecies interactions and species—phenotype relations such as competitive and cooperative interactions leveraging to big microbiome data. These endeavors have facilitated the discovery of previously unknown principles of microbial world and expedited the understanding of the disease mechanism. In this review, we will summarize current computational efforts in microbiome data mining for discovering microbial interactions and relationships including dimension reduction and data visualization, association analysis, microbial network reconstruction, as well as dynamic modeling and simulations.

1 Introduction

There are two fundamental research questions in microbial ecology [1]. The first question is how many kinds of microbes are in a microbial environment? It is estimated that there are about 1000 different kinds of bacteria in a human gut; however, it is very hard to know what kind of bacteria exist there [2]. The second

X. Jiang · X. Hu (✉)
School of Computer, Central China Normal University,
Wuhan 430079, Hubei, China
e-mail: xh29@drexel.edu

X. Jiang
e-mail: xpjiang@mail.ccnu.edu.cn

X. Hu
College of Computing and Informatics,
Drexel University, Philadelphia, PA 19104, USA

© Springer India 2016   221
S. Pyne et al. (eds.), *Big Data Analytics*, DOI 10.1007/978-81-322-3628-3_12

question is what kind of metabolic functions are the major ones and how microbes collaborate to complete a complex biological function? To answer these fundamental questions, we need to study the structure and the function of genes and microbes in a microbial community.

Metagenomic sequencing that investigate the overall DNA content of a microbial community by high-throughput sequencing technologies without culture any individual species has become a powerful tool for microbiome research [3]. It allows researchers to characterize the composition and variation of species across environmental samples, and to accumulate a huge amount of data which make it feasible to infer the complex principle of microbial ecology (Fig. 1).

Microbiome sequencing can be used to investigate not only the diversity of microorganisms but also that of gene functions. For example, functional profiles including metabolic profiles [5], protein family profile [6], or taxonomic profiles which include operation taxonomic unit table [7], genus, and species profile [8]. These profiles indicate the abundance of functional or taxonomic categorizations in a microbiome sample which can be represented by a matrix or a tensor. In a profile matrix, entries indicate the abundance of a function or taxa (See Fig. 2). Investigating the profile can help us to answer the two fundamental questions. Interdisciplinary methods from mathematics, statistics, and pattern recognition could help to identify the hidden structure and patterns embedded in a microbial community and discover nonlinear relationships between microorganisms and environmental or physiological variables (Fig. 2).

To state the two fundamental questions in microbial community clearly, we also have to know how different species coexist and interact in a natural environment. Microbiology is experiencing a "holistic thinking" trend that emphasizes the interactions of different elements in a microbial community using network analysis and modeling tools (at the bottom of Fig. 2). Microbial interaction network (MIN, e.g., species–species interaction network) shapes the structure of a microbial community and hence forms its ecosystem function and principle such as the regulation of microbe-mediated biogeochemical processes. Deciphering interspecies interaction is challenging by wet lab due to the difficulty of co-culture experiments and the complicate types of species interactions. Researchers have started to infer pairwise interspecies interactions such as competitive and cooperative interactions leveraging to large-scale microbiome data including metagenomes [2, 9, 10], microbial genomes [11–13], and literature data [14]. These efforts have facilitated the discovery of previously unknown principles of MIN, verified the consistency, and resolved the contradiction of the application of macroscopic ecological theory in microscopic ecology.

In this review, we will investigate computational studies in big microbiome data mining. We will emphasize computational methods in a unit framework as shown in Fig. 2 which are related to dimension reduction and data visualization as well as microbial network construction and simulation.

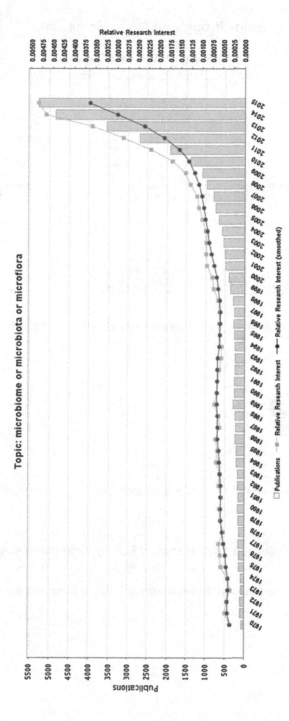

Fig. 1 The rapid extension of microbiome studies. The figure is made by the online tool GoPubMed [4], using key words "microbiome or microbiota or microflora." The relative research interest relates publication statistics to the overall number of scientific publications indexed by PubMed which provides a better view of research activity and importance than an absolute number of papers. The smoothed relative research interests are the relative research interest smoothed by a sliding window of 5 years. [4]

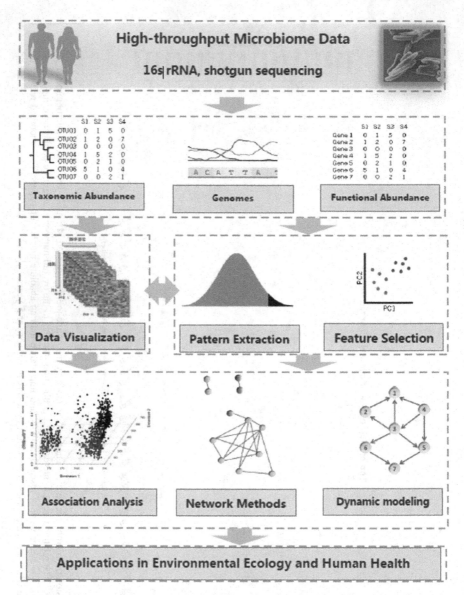

Fig. 2 A data mining framework for identifying and analyzing interactions and relationship from microbiome data

2 Dimension Reduction and Visualization of Microbiome Data

2.1 Dimension Reduction and Data Visualization of Microbiome Profiles

Dimension reduction and data visualization converts the high-dimensional dataset $X = \{x_1, x_2, \ldots, x_n\}$ into two- or three-dimensional data $Y = \{y_1, y_2, \ldots, y_n\}$ that can be displayed in a scatter plot at a computer screen for exploring data conveniently. The study of microbial ecosystem biology, some classical statistical, and pattern recognition methods for data dimension reduction and visualization has been applied to analyze microbiomic data [1], such as principal component analysis (PCA) [15], multidimensional scaling analysis (MDS) [16], nonnegative matrix factorization (NMF) [6, 17], and probabilistic topic model (PTM) [18], (we will discuss PTM in detail in the next subsection). Arumugam et al. [19] found that the type of microbial compositions in a human gut is not random and it can be classified into at least three enterotypes by integrating PCA and clustering analysis. The mechanism of enterotypes is still unknown but it may relate the action of human immunity systems to bad or good bacteria, or to different methods of excrete wastes from the body. Similar to blood type, enterotype is not related to demographic factors such as country, age, gender, race, or other body index. Metric-MDS is one subclass of MDS, which is also called principal coordinate analysis (PCoA) [19]. After combining with phylogenetic distance, PCoA is widely used in the visualization and interpretation of microbiomic data [20–26]. For instance, the integration of PCoA with unweighted UniFrac distance has been used in analyzing the age and geographic distributions of 531 human microbiomic [20].

NMF framework has been used in analyzing metagenomic profiles to gain a different and complementary perspective on the relationships among functions, environment, and biogeography of global ocean [6, 17]. To investigate if the metagenomic dataset is better modeled by manifold rather than Euclidean space, an analysis framework based on a nonlinear dimension reduction method—Isometric Feature Mapping (Isomap)—has been proposed to explore the nonlinear structure in metagenomic profiles [27]. In contrast to PCA and NMF, Isomap identified nonlinear structure embedded in the data clearly. The results indicate that the nonlinear components are significantly connected to environmental factors.

2.2 Probabilistic Modeling of Microbiome Data

Probabilistic topic models (PTM) can be used to build a generative model to simulate the generation process of a microbial community based on the taxonomic or functional categorization from metagenomes. In this model, a microbial community can be regarded as a probabilistic mixing of microbial "topics" rather than a linear combination of "components." These methods provide a generative mechanism of microbial community which is helpful in understanding the generating of complex microbial ecology.

Based on the functional elements derived from non-redundant CDs catalog in human gut microbiome, the configuration of functional groups in metagenome samples can be inferred by probabilistic topic modeling [18, 28]. The generative model assumes that a microbiome sample could be generated in a style of writing a document. Each microbial sample (assuming that relative abundances of functional elements are already obtained by a homology-based approach) can be considered as a "document," which is a mixture of functional groups, while each functional group (also known as a "latent topic") is a weighted mixture of functional elements (including taxonomic levels, and indicators of gene orthologous groups and KEGG pathway mappings). The functional elements bear an analogy with "words". Estimating the probabilistic topic model can uncover the configuration of functional groups (the latent topic) in each sample. The results are consistent with the recent discoveries in fecal microbiota study of inflammatory bowel disease (IBD) patients [2], which demonstrate the effectiveness of probabilistic topic model.

Another method used Dirichlet multinomial mixture distributions to model microbial community [29] which provides model-based alternatives for both clustering and classification of microbial communities. The success of probabilistic models is not surprising because of the probability distribution nature of microbiome profiles, i.e., the relative abundance profile is a composition of probability distribution.

3 Large-Scale Association Analysis and Microbial Network Reconstruction

Network methods have great potential to discover biological characteristics associated with human health, disease, and environment change. Microbiomics data usually has high dimensionality characteristics, namely the number of functional or taxonomic classifications is far greater than the number of the number of samples (machine learning often referred to as the curse of dimensionality). Algorithms including correlation analysis, statistical tests, and complex network approach must be corrected or improved in order to obtain a reasonable application in microbiome data. Furthermore, microbiomics data has its complex data specialty that provides difficulties for current methods to identify true relationships.

3.1 Relationship Between Microbiomic Data and Metadata

Linear correlation or regression methods are also employed to investigate the relationships among taxa or functions and their relationships to existing environmental or physiological data (metadata), such as Pearson correlation and Canonical correlations analysis (CCA). CCA has been proposed for investigating the linear relationships of environmental factors and functional categorizations in global ocean [15, 30, 31]. Canonical correlation analysis (CCA) explores linear correlation between the data of the principle dimensions by maximizing the microbial and environmental variables, in order to find the most relevant environmental variables and functional units.

Generally speaking, in the practical application facing the curse of dimensionality problem, the direct application of CCA is not feasible; we must first apply dimensionality reduction and then use the CCA or the regularization of CCA. Literature [30] provided a quantitative analysis of the microbial metabolic pathways in global marine environment by CCA, and investigated the environmental adaptability to find related metabolic markers (called "CCA fingerprint" in the paper). Literature [31] investigated metagenomic features using more environment variables, climatic factors, and the relationship between primary productions through CCA. It was found that temperature, light, and other climatic factors determine the molecular composition of marine biological factors.

3.2 Microbial Interaction Inference

The interaction among species is very important to an ecological system function, for example, the structure of food web plays a pivotal role in the evolution of the macro-ecological environment. Metagenomic and 16S sequencing allows inference of large-scale interactions between microbial species [32]. In recent years, computational methods for microbial network inference are proposed based on the co-occurrence pattern or correlations of microbial species group. For instance, similarity, relevance and mutual information method can be used to predict the simple pairwise species interactions [7]; while regression and rule-based data mining methods can be used to predict more complex species interactions.

After the completion of the interaction prediction, network analysis methods will help deep understanding of the microbial structure impact on the ecological environment of the visualization of species interaction network, the extraction of complex structures, the structural analysis of network attributes (such as distribution of the average path length and modularity, etc.) and the exploratory discovery of the relationship between the environment and microbes.

Reshef et al. [33] proposed a new nonparametric statistical method used to extract the complex relationship between large-scale data, which is based on "Maximum Information Coefficient" (MIC) without using prior assumption of function types, nonparametric optimization can be found through a complex association between variables. MIC is based on the idea that if two variables have a relationship, then there should exist a scatter plot mesh of the two variables that make most of relationships of the data points concentrated in a few cells of the grid medium. By searching for this "optimal" grid computing, MIC can discover the statistical association between variables, and the statistical framework based on MIC is called MINE (Maximal information-based nonparametric exploration). Reshef et al. compared MINE with other methods and found MINE is more suitable for large-scale data on the association mining with fast computing speed. MIC has attracted widespread interest and has been successfully applied to the high-throughput microbiomics data analysis [34–37].

Chaffron et al. used Fisher's exact test with FDR correction of p value to build a large-scale network in a scale of microbial community based on co-occurrence patterns of microorganism; many microorganisms associated in network are also similar evolutionarily [38]. Zupancic and others used Spearman correlation coefficient for constructing a network of relationships between intestinal bacteria from Amish (Amish) people and used regression analysis to investigate the relationships between three bacterial networks (corresponding to the three enterotypes) and metabolic phenotypes; they found 22 bacterial species and four OTU related to metabolic syndrome [39].

Faust et al. analyzed the data from the first stage of the Human Microbiome Project–16S rRNA sequence data of 239 individuals with 18 body positions (totally 4302 samples) to build a global interaction network of microorganisms [40]. The method for building interaction network in which the introduction of a extended generalized linear models (Generalized Boosted Linear Models, GBLM) models the prediction of the target position (target site, ts) with species taxa (target taxa, tt) abundance $X_{tt,ts}$ by the source position (source site, ss) with species taxa (source taxa, st) abundance $X_{st,ss}$:

$$X_{st,ss} : X_{tt,ts} = \overline{X}_{st,ss} + \sum_{st} \beta_{tt,ts,st,ss} X_{st,ss}. \tag{1}$$

Also due to the high-dimensional nature of the data, the model is prone to over-fitting. Hence, a modified linear model (Boosted Linear Models) is proposed, based on the linear regression analysis model–LASSO [41]. The final interaction network integrates GBLM and other correlation or similarity measure (Pearson correlation, Spearman correlation coefficient, Kullback–Leibler divergence, Bray–Curtis distance) resulting species interaction networks. The results showed that the co-occurrence of microbial interactions often occur in similar parts of the body, which was in line with "the different parts of the body are different niche" hypothesis, while interacting with the exclusion often occurs in between evolutionarily distant species, so tend competition [41].

3.3 Network Reconstruction for Compositional Data

As mentioned earlier, the taxonomic profile has another very significant feature that it belongs to a special data types—composition data [42–44]. Due to the small number of species in a larger number of taxa exist, a lot of negative correlation are generated by bias in the application of Pearson correlation coefficient on conventional compositional data, and the effect is known as "Composition effect." To overcome this problem, Friedman et al. systematically investigate the species (or functional) diversity effect on the compositional data and found that when the data has a very low diversity of species, the composition effect can be ignored, and vice versa. Therefore, although the same issues in gene expression data by high-throughput sequencing, there is not a number of dominant genes, and thereby the compositional effect in data of DNA chips, RNA-seq or ChIP analysis-seq is not significant even negligible. Friedman proposed a new computational framework (SparCC) to overcome the compositional effect [45].

SparCC transforms the original matrix OTU relative to abundance ratio x_i by

$$y_{ij} = log\frac{x_i}{x_j} = logx_i - logx_j \tag{2}$$

to introduce a new variable y_{ij}. New variable y_{ij} has several good statistical properties than x_i. The results found that SparCC can extract relationships from data with high accuracy. Experimental results of SparCC on HMP data show that the relationships obtained by Pearson correlations have about 3/4 wrong, and it is estimated that about 2/3 of the true relations have not been identified by Pearson correlation. This result suggests that a deep understanding of the data can inspire a new set of data analysis methods for microbiological analysis and there are still a lot of space and challenges for dealing with microbiomic data.

3.4 Inferring Species Interaction from Time Series Data

Microbial abundance dynamics along the time axis can be used to explore complex interactions among microorganisms. It is important to use time series data for understanding the structure and function of a microbial community and its dynamic characteristics with the perturbations of the external environment and physiology. Current studies confined to use time sequence similarity, or clustering time series data for discovering dynamic microbial interactions; these methods do not take full advantage of the time sequences. We have proposed an extension of VAR models for species interaction inference [46].

Again, due to the high-dimensional nature of microbiomics data, the number of samples is far greater than the number of microorganisms; direct interaction inference by VAR is not feasible. In our previous studies, we have designed a graph Laplacian regularization VAR (GVAR) method for analyzing the human microbiome [46, 47]. We validated the strength on several datasets. We found that our approach improves the modeling performance significantly. The experimental results indicate that GVAR achieves better performance than the sparse VAR model based on elastic net regularization [46, 47].

3.5 Data Integration for Inferring Species Interactions

Data integration will take the extracted interactions and relations from heterogeneous datasets and identify the knowledge gaps in the different knowledge sources. The input for the data integration module is interactions (that can be extracted from any inferring methods or pipeline), and the output is knowledge that has been validated and deemed consistent with existing knowledge. Figure 3 shows a proposed hierarchical structure in a data integration process. Microbial interactions in genus or species level will be inferred depending on the availability of data.

Using data integration, we can also collect other entities information (e.g., species–attributes) and relations (e.g., species–environment and species–disease relation [48]) because they are important for biological correlation discovery. Statistical machine learning approaches for consistency checking of inferred interactions and relations from various resources, validation, and qualitative reasoning on integrated knowledge graph will be implemented. The integrated data could provide a comprehensive and hierarchical knowledge graph for many kinds of relationships in microbial community study.

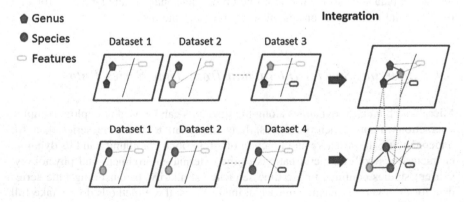

Fig. 3 The integration of inferred microbial interactions and relations in a hierarchical structure

4 Large-Scale Dynamics Simulation for Microbial Community

To better understand the biological phenomena, it is necessary to study the time evolution of microbial interaction to get in-depth understanding of a single microbial community. Dynamics analysis for the evolution of microbial interactions can reveal the mechanisms and rules of microbial ecosystem. Literature [49] analyzed the human intestinal antibiotics trends and revealed the effects of antibiotics on intestinal microbial community diversity; Literature [34] studied the dynamics in intestinal microbial community of a developing infant. However, microbial system is a large system which is hard to simulate. For example, it is estimated that in the microbial ecosystem of human gut, there are about 1000 kinds of bacteria with a billion of bacteria and more than 5 million genes in more than 6000 orthologous gene family.

4.1 Dynamic Models for Microbial Interaction Network

For discovery of the dynamics of microbial communities (Dynamic signature), Gerber et al. developed a computational framework based on Bayesian methods [50], and apply it to a time series gut microbiomic data under antibiotic treat [49]. In this method, OTUs are assigned to a few time series prototypes feature (prototype signature) by a probability distribution. This identifies which OUTs are related to antibiotics reaction. Mounier and co-workers using the generalized gLV (Generalized Lotka–Volterra) equations to simulate the interaction of microorganisms in fermentation yeast of cheese [51].

They predicted three negative interactions between species which have been experimentally confirmed. gLV equation is a prey–predator model, the so-called "destroy winner" model that can be used in this proposal to simulate the micro-scale microbial dynamics. Assuming an environment with N microorganisms, the growth rate b_i of microorganism X_i, the interaction strength a_{ij} between X_i and X_j, consider the following promotion LV equation

$$X_i(t) = X_i(t)\left(b_i + \sum_{j=1}^{N} a_{ij}X_j(t)\right) \tag{3}$$

where $X_i(t)$ is the abundance profile of the species i at time t. GLV equations can be used to investigate the stability and chaotic behavior of a microbial environment. This model can also analyze the robustness of an ecosystem through the investigation that the strength of an outside interference on a microbial environment in order to determine whether the impact of outside interference reversible.

In microbial environment, predator is usually a virus or winner bacteria in food competent. An increase in the abundance of microbial species also increases the probability of its prey by the virus (winner bacteria), making the system stable at large scale but with rapid fluctuations at small scale. Hoffmann and collaborators used a modified gLV model to simulate the abundance of marine phage [52] and found that the phage–host community dynamics is in transition between long-term low abundance and short prosperous. The model predicts that the abundance of phage–host prosperous period in its moment of its risk period (coincides to "destroy winner" model), and then the phage abundance decreased rapidly, after that slows down at low levels of abundance to survive until the host arrive the next heyday.

4.2 Dynamic Modeling for Metabolic Network at Community Level

Flux balance analysis (Flux Balance Analysis, FBA) proposed in 1999 is a mathematical model for metabolic network at genome scale [53]. Compared with traditional simulation methods, FBA model has satisfied computing efficiency. FBA can be calculated to estimate the metabolic flux steady state within a few seconds for moderate scale metabolic networks (>2000 responses).

The basic idea of FBA is as follows: Consider the predicted maximum yield of a target substance while the organisms reach a steady state, and the overall metabolic flux distribution within the network should be balanced. FBA can effectively use the information of stoichiometric matrix and integrate constraint conditions to find the optimal solution of a simulation system [53]. According to the flux of each reaction and stoichiometric coefficients, all metabolite quality equations constitute simple ordinary differential equations. Set the stoichiometry matrix as S, the amount of metabolic flow v, and the system reaches steady state when

$$S \cdot v = 0. \tag{4}$$

Stolyar first analyzed the multi-species metabolic model [54]; the model has two simple microbes—*Desulfovibrio vulgaris* and *Methanococcus maripaludis*. FBA analysis on the simple bacterial community accurately predicted the growth of a number of co-culture features, including the ratios of the two species. It suggested that the metabolic network reconstruction and stoichiometric analysis can not only predict the metabolic flow and growth phenotype for single species but also for microbial communities.

117 genome-scale metabolic network models were published by Feb 2013. In microbial ecosystem biology, multi-species FBA will also show its potential. Constraint-based modeling (CBM) was already used for the inference of three potential interactions [13]: negative, where two species compete for shared resources; positive, where metabolites produced by one species are consumed by another producing a synergistic co-growth benefit; and neutral, where co-growth has no net effect. In a two-species system, the CBM solver aims to explore the type of interactions by comparing the total biomass production rate (denoted AB) in the pairwise system to the sum of corresponding individual rates recorded in their

Fig. 4 A constraint-based modeling to model pairwise interaction

$$v_{BM,AB} = max \sum_{m \in \{A,B\}} v_{BM,m}$$

Subject to:

$$SV = 0$$

$$v_{i,min} \leq v_i \leq v_{i,max}$$

$$v_i \in V$$

individual growth (denoted A + B). The CBM model is defined in Fig. 4, where $v_{BM,m}$ is the maximal biomass production rate in a system m, corresponding to species A and B. When AB $\ll$ A + B, A and B have a competitive relationship.

5 Conclusion

To comprehensively explore microbial ecosystems and make the most potential of public microbiomic data, it is necessary to integrate different types of data in different scales and develop novel computational methods to overcome these assumptions and consider the microbiomic data properties in the analysis procedure. Data mining for big microbiome data helps us to discover hidden structures and patterns in microbiomic data, to understand the structure and function of microbial community, and the interaction between microbial community and its host environment. The development of these methods needs interdisciplinary efforts in computer science, mathematics, and biology by integrating high-throughput sequencing data such as metagenome, 16 s rRNA sequencing data, metatranscriptome, meta-metabolism, and high-quality metadata, such as environmental, chemical, physical, and geographic data and other ecological, genomics data.

Acknowledgments This work was supported in part by NSF IIP 1160960, NNS IIP 1332024, NSFC 61532008, and China National 12-5 plan 2012BAK24B01 and the international cooperation project of Hubei Province (No. 2014BHE0017) and the Self-determined Research Funds of CCNU from the Colleges' Basic Research and Operation of MOE (No. CCNU16KFY04).

References

1. Wooley JC, Godzik A, Friedberg I (2010) A primer on metagenomics. PLoS Comput Biol 6 (2):e1000667
2. Qin J et al (2010) A human gut microbial gene catalogue established by metagenomic sequencing. Nature 464(7285):59–65
3. Cho I et al (2012) Antibiotics in early life alter the murine colonic microbiome and adiposity. Nature 488(7413):621–626
4. Lin J, Wilbur WJ (2007) PubMed related articles: a probabilistic topic-based model for content similarity. BMC Bioinform 8

5. Abubucker S et al (2012) Metabolic reconstruction for metagenomic data and its application to the human microbiome. PLoS Comput Biol 8(6):e1002358
6. Jiang X et al (2012) Functional biogeography of ocean microbes revealed through non-negative matrix factorization. PLoS ONE 7(9):e43866
7. Karlsson FH et al (2013) Gut metagenome in European women with normal, impaired and diabetic glucose control. Nature 498(7452):99–103
8. Morgan, X.C. and C. Huttenhower, *Chapter 12: Human microbiome analysis.* PLoS Comput Biol, 2012. **8**(12): p. e1002808
9. Ren TT et al (2013) 16S rRNA survey revealed complex bacterial communities and evidence of bacterial interference on human adenoids. Environ Microbiol 15(2):535–547
10. Chaffron S et al (2010) A global network of coexisting microbes from environmental and whole-genome sequence data. Genome Res 20(7):947–959
11. Carr R, Borenstein E (2012) NetSeed: a network-based reverse-ecology tool for calculating the metabolic interface of an organism with its environment. Bioinformatics 28(5):734–735
12. Greenblum S et al (2013) Towards a predictive systems-level model of the human microbiome: progress, challenges, and opportunities. Curr Opin Biotechnol 24(4):810–820
13. Shoaie, S., et al., *Understanding the interactions between bacteria in the human gut through metabolic modeling.* Scientific Reports, 2013. **3**
14. Freilich S et al (2010) The large-scale organization of the bacterial network of ecological co-occurrence interactions. Nucleic Acids Res 38(12):3857–3868
15. Patel PV et al (2010) Analysis of membrane proteins in metagenomics: Networks of correlated environmental features and protein families. Genome Res 20(7):960–971
16. Temperton B et al (2011) Novel analysis of oceanic surface water metagenomes suggests importance of polyphosphate metabolism in oligotrophic environments. PLoS ONE 6(1): e16499
17. Jiang X, Weitz JS, Dushoff J (2012) A non-negative matrix factorization framework for identifying modular patterns in metagenomic profile data. J Math Biol 64(4):697–711
18. Chen X et al (2012) Estimating functional groups in human gut microbiome with probabilistic topic models. IEEE Trans Nanobiosci 11(3):203–215
19. Arumugam M et al (2011) Enterotypes of the human gut microbiome. Nature 473 (7346):174–180
20. Yatsunenko T et al (2012) Human gut microbiome viewed across age and geography. Nature 486(7402):222–227
21. Wu GD et al (2011) Linking long-term dietary patterns with gut microbial enterotypes. Science 334(6052):105–108
22. Hildebrand F et al (2013) Inflammation-associated enterotypes, host genotype, cage and inter-individual effects drive gut microbiota variation in common laboratory mice. Genome Biol 14(1):R4
23. Koren O et al (2013) A guide to enterotypes across the human body: meta-analysis of microbial community structures in human microbiome datasets. PLoS Comput Biol 9(1): e1002863
24. Moeller AH et al (2012) Chimpanzees and humans harbour compositionally similar gut enterotypes. Nat Commun 3:1179
25. Jeffery IB et al (2012) Categorization of the gut microbiota: enterotypes or gradients? Nat Rev Microbiol 10(9):591–592
26. Siezen RJ, Kleerebezem M (2011) The human gut microbiome: are we our enterotypes? Microb Biotechnol 4(5):550–553
27. Jiang X et al (2012) Manifold learning reveals nonlinear structure in metagenomic profiles. In: IEEE BIBM 2012
28. Chen X et al (2012) Exploiting the functional and taxonomic structure of genomic data by probabilistic topic modeling. IEEE-ACM Trans Comput Biol Bioinform 9(4):980–991
29. Holmes I, Harris K, Quince C (2012) Dirichlet multinomial mixtures: generative models for microbial metagenomics. Plos ONE 7(2)

30. Gianoulis TA et al (2009) Quantifying environmental adaptation of metabolic pathways in metagenomics. Proc Natl Acad Sci USA 106(5):1374–1379
31. Raes J et al (2011) Toward molecular trait-based ecology through integration of biogeochemical, geographical and metagenomic data. Mol Syst Biol 7:473
32. Friedman J, Alm EJ (2012) Inferring correlation networks from genomic survey data. PLoS Comput Biol 8(9):e1002687
33. Reshef DN et al (2011) Detecting novel associations in large data sets. Science 334 (6062):1518–1524
34. Koren O et al (2012) Host remodeling of the gut microbiome and metabolic changes during pregnancy. Cell 150(3):470–480
35. Anderson MJ et al (2003) Biochemical and toxicopathic biomarkers assessed in smallmouth bass recovered from a polychlorinated biphenyl-contaminated river. Biomarkers 8(5):371–393
36. Hinton D et al (2003) 'Hit by the wind' and temperature-shift panic among Vietnamese refugees. Transcult Psychiatry 40(3):342–376
37. Kamita SG et al (2003) Juvenile hormone (JH) esterase: why are you so JH specific? Insect Biochem Mol Biol 33(12):1261–1273
38. Chaffron S et al (2010) A global network of coexisting microbes from environmental and whole-genome sequence data. Genome Res 20(7):947–959
39. Zupancic M et al (2012) Analysis of the gut microbiota in the old order amish and its relation to the metabolic syndrome. PLoS ONE 7(8):e43052
40. Faust K et al (2012) Microbial co-occurrence relationships in the human microbiome. Plos Comput Biol 8(7)
41. Lockhart R et al (2014) A significance test for the Lasso. Ann Stat 42(2):413–468
42. Negi JS et al (2013) Development of solid lipid nanoparticles (SLNs) of lopinavir using hot self nano-emulsification (SNE) technique. Eur J Pharm Sci 48(1–2):231–239
43. Xie B et al (2011) m-SNE: multiview stochastic neighbor embedding. IEEE Trans Syst Man Cybern B Cybern
44. Greene G (2010) SNE: a place where research and practice meet. J Nutr Educ Behav 42(4):215
45. Friedman J, Alm EJ (2012) Inferring correlation networks from genomic survey data. Plos Comput Biol 8(9)
46. Jiang X et al (2014) Predicting microbial interactions using vector autoregressive model with graph regularization. IEEE/ACM Trans Comput Biol Bioinform (in press). doi:10.1109/TCBB.2014.2338298
47. Jiang X et al (2013) Inference of microbial interactions from time series data using vector autoregression model. In 2013 IEEE International conference on bioinformatics and biomedicine (BIBM). IEEE
48. Ishak N et al (2014) There is a specific response to pH by isolates of Haemophilus influenzae and this has a direct influence on biofilm formation. BMC Microbiol 14:47
49. Dethlefsen L, Relman DA (2011) Incomplete recovery and individualized responses of the human distal gut microbiota to repeated antibiotic perturbation. Proc Natl Acad Sci USA 108 (Suppl 1):4554–4561
50. Gerber GK (2014) The dynamic microbiome. FEBS Lett
51. Mounier J et al (2008) Microbial interactions within a cheese microbial community. Appl Environ Microbiol 74(1):172–181
52. Hoffmann KH et al (2007) Power law rank-abundance models for marine phage communities. FEMS Microbiol Lett 273(2):224–228
53. Orth JD, Thiele I, Palsson BO (2010) What is flux balance analysis? Nat Biotechnol 28 (3):245–248
54. Stolyar S et al (2007) Metabolic modeling of a mutualistic microbial community. Mol Syst Biol 3(1):92

A Nonlinear Technique for Analysis of Big Data in Neuroscience

Koel Das and Zoran Nenadic

Abstract Recent technological advances have paved the way for big data analysis in the field of neuroscience. Machine learning techniques can be used effectively to explore the relationship between large-scale neural and behavorial data. In this chapter, we present a computationally efficient nonlinear technique which can be used for big data analysis. We demonstrate the efficacy of our method in the context of brain computer interface. Our technique is piecewise linear and computationally inexpensive and can be used as an analysis tool to explore any generic big data.

1 Big Data in Neuroscience

In the last decade, big data has impacted the way research is carried out in areas including genomics [37, 45], proteomics [34, 44] and physics [17]. Neuroscience is a late entrant in the field of big-data analysis, and a large number of neuroscience studies still involves hypothesis-driven science using small-scale data. However, with the introduction of big-data projects like the BRAIN initiative [25] in the United States or the Human Brain Mapping initiative [36] in Europe, there has been a gradual transition toward big data collection, analysis, and interpretation [1].

Technological advances have made it possible to simultaneously record big neural data and behavioral measurements, while machine learning and data mining algorithms are being routinely used to explore the relationship between these data. Because of their simplicity and ability to be executed in real time, statistical techniques based on linear, parametric models are often used for this purpose. In the face of large data dimension, many of these methods are hindered by the small sample size problem and curse of dimensionality. Nonlinear and/or nonparametric techniques are

K. Das (✉)
Department of Mathematics and Statistics, IISER Kolkata, Mohanpur 741246, India
e-mail: koel.das@iiserkol.ac.in

Z. Nenadic
Department of Biomedical Engineering, University of California,
Irvine, CA 92697, USA

© Springer India 2016
S. Pyne et al. (eds.), *Big Data Analytics*, DOI 10.1007/978-81-322-3628-3_13

a viable alternative, although their computational cost may be prohibitive for real-time implementation. In this chapter, we present a nonlinear technique suitable for analysis of big data in neuroscience. The technique retains computational simplicity of linear methods, and is therefore amenable to real-time implementations.

One of the key steps in performing multivariate analysis of big data is dimensionality reduction. In this chapter, we focus on the key aspect of computational challenges and how dimensionality reduction techniques can be potentially used to explore, analyze, and interpret results. More specifically, we analyze the challenges posed by high-dimensional neural signals and describe a particular feature extraction technique which is well suited for multivariate analysis of big data. We conclude by demonstrating the efficacy of the method to decode neural signals in a brain-computer interface (BCI) setup.

Brain–computer interface is a system capable of enabling the interaction of a user with its environment without generating any motor output [48, 72]. It may be envisioned that future BCI systems will gain a widespread adoption in the fields of neuroprosthesis and neurorehabilitation. Researchers have proposed designing cortical prosthetic devices to aid the millions of people suffering from paralysis, due to spinal cord injury (SCI), strokes, or neurological conditions such as Lou Gehrig's disease, cerebral palsy, and multiple sclerosis. The ultimate goal of a BCI-controlled prosthetic system is to decode in real time information contained in neural signals, and to use this information for control of assistive devices, such as arm [21, 68] or leg [14, 29] prostheses. In addition, BCI systems have been explored as a means for delivering neuro-rehabilitative therapy to those affected by chronic stroke [8, 15, 38, 49].

Typically, the decoding is facilitated by recording neural correlates of the intended task (e.g., movement intentions or movement trajectories), and by accumulating sufficient data over multiple trials into a *training database*. Future data can then be compared against the training database, and the intended task can be decoded. Although this article focuses on analyzing big neural data specifically in the context of BCIs, the method described here can be potentially useful for a wide range of applications relying on big data.

1.1 Dimensionality Reduction in Neural Signals

Neural signals are often high dimensional and therefore require special handling including dimensionality reduction followed by feature extraction. When approached from linear parametric model standpoint, statistical analysis of high dimensional, spatio-temporal neural signals is impeded by the small sample size problem and curse of dimensionality.

1. **Small Smaple Size Problem**.

   This problem arises when data dimension, n, exceeds the number of samples, N. Under these conditions, [19], the data space becomes sparse resulting in sample statistics being inaccurate [19, pp. 70], [24]. Consequently, the covariance matrices become highly singular, and a straightforward implementation of classical techniques such as linear discriminant analysis (LDA) [18], is ruled out.

2. **Curse of Dimensionality**.

The storage and manipulation of big datasets poses a severe computational challenge. For example, second-order-statistics-based methods give rise to large covariance matrices, whose inversion and spectral decomposition may become infeasible even with state-of-the-art computer architectures.

Efficient feature extraction methods are often employed to circumvent the above-mentioned hurdles, whereby low-dimensional informative features are extracted from the big data and irrelevant data are discarded as noise. The curse of dimensionality and small sample size problem are typically mitigated in the reduced feature space. Several feature extraction techniques have been proposed including classical discriminant analysis methods [18, 51] and their variants [35, 69, 76] and probabilistic measure-based techniques [42, 55]. However, most of the proposed schemes cannot resolve the small sample size problem and/or the curse of dimensionality. Shrinkage-based discriminant feature extraction (DFE) techniques [9, 56] can effectively solve the small sample size problem by making the covariance matrix regular, but they do not address the problem of *curse of dimensionality*. Additionally, the shrinkage parameter needs to be estimated through expensive procedures [22, 63]. Subspace-based decomposition techniques [3, 6, 67, 69, 76, 78] provide alternate solution to address the problem of high-dimensional data by discarding noninformative subspaces. The most widely used subspace-based approach is to reduce the data dimension via principal component analysis (PCA). Kirby and Sirovich [30] were perhaps the first researchers to use PCA in the context of human face detection. The other notable early approaches using PCA-based data reduction are the eigenface technique developed by Turk and Pentland [67] and Fisherface method proposed by Belhumeur et al. [3]. Many LDA-based approaches [6, 23, 69, 76–78] have been proposed in the last decade which address the small sample size problem and curse of dimensionality and are especially successful in the context of face recognition. Several kernel-based techniques [57, 75] and manifold-based learning techniques [20, 53] have also been used efficiently in many applications suffering from small sample size problem. However, most of these methods were proposed for applications like face/object recognition and when applied directly to high-dimensional, spatio-temporal neural data, they did not produce satisfactory results [10, 11]. In addition, kernel-based methods are typically computationally demanding, and may not conform to real-time processing demands. Hence, the features in neuroscience have been traditionally extracted based on heuristic criteria.

A common heuristic approach is to separately process the data spatially and temporally. Data can be analyzed spatially [39, 47, 74] by applying the Laplacian filter, followed by temporal processing. The inherent assumption in this approach is that space and time are separable, which is likely to be violated in practice giving rise to suboptimal performance.

Temporal processing on the other hand involves spectral decomposition of brain data and the utilization of power in various frequency bands (e.g., μ-band or β-

band [47, 74]). In some applications, such approach could be physically intuitive [52], but these ad hoc spectral features produce suboptimal classification of brain signals. Furthermore, it could give rise to significant individual difference since some users are unable to control the μ and β-rhythms [33]. Ranking individual features [59, 60, 66], or even electrodes [52], according to their individual usefulness, and constructing feature vectors by concatenating few of the most dominant individual features is another popular strategy used in conjunction with the above approaches. The major drawback of this approach is that the joint statistical properties of features are ignored [7, 31, 65], which may produce suboptimal feature vectors. Elaborate algorithms for the concatenation of features exist [19], but they are combinatorially complex [7] and have a limited practical applicability. Another approach to dimensionality reduction is to perform temporal binning of data, i.e., replace data over a suitably chosen bin with their time-averaged value. This technique is often used in conjunction with traditional feature extraction approaches [43, 52]. An obvious weakness of this technique is that some information is inevitably lost by binning. Furthermore, the estimation of the optimal bin size is a challenging problem and renders this method unappealing. Recently data-driven adaptive approaches like common spatial pattern (CSP) filtering and its variants [2, 5, 26, 40, 50, 70] have been used successfully in many BCI applications. The number of channels used is reduced systematically in CSP and this kind of approach has been efficiently used in motor imagery-based studies. However, spatial and temporal processing remain separated in CSP and joint extraction of spatial and temporal features is not possible.

In the next section,a principal component analysis-based feature extraction/ classification scheme is described. The feature extraction method is capable of dealing with sparse high-dimensional neural data. The method yields a piecewise linear feature subspace and is particularly well suited to hard recognition problems where classes are highly overlapped, or in cases where a prominent curvature in data renders a projection onto a single linear subspace inadequate. Unlike state-of-the-art face recognition techniques, the method performs reasonably well on a variety of real-world datasets, ranging from face recognition to bioniformatics and brain imaging.

2 Classwise Principal Component Analysis

We develop a computationally efficient, locally adaptable feature extraction and classification technique which can handle big data suffering from the small sample size conditions. Our method uses PCA to reduce data dimension globally, while preserving the class-specific discriminatory information to facilitate subsequent classification. The technique is based on a classwise PCA (CPCA) wherein we reject noninformative subspace in the high-dimensional data by using PCA on each class. The resulting low-dimensional feature space retains class-specific information and is used for classification purpose.

Fig. 1 2D examples for binary classes where PCA and LDA does not work. The *black straight lines* indicate optimal subspace by our CPCA-based method. LDA and PCA subspace are denoted in *gray dashed dot lines*. **a** illustrates Gaussian classes with overlapped means. **b** shows bimodal uniform classes. All the classes have equal priors, $P(\omega_1) = P(\omega_2)$

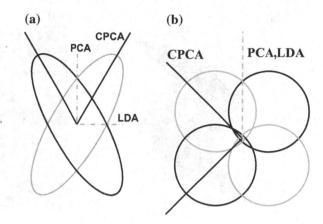

Figure 1 illustrates a couple of two dimensional (2D) binary class examples where the classical PCA and LDA fail, however our technique is able to capture the discriminatory information. Monte Carlo integration validates our finding; the Bayes error is about 40 % using PCA and LDA and 24 % using our method in Fig. 1a, and 50 % and 27 % in Fig. 1b, respectively.

2.1 Algorithm

In general, for a c-class problem, our technique gives rise to a family of c subspaces $\{S_1, S_2, \ldots, S_c\}$, so that if data is confined to a low-dimensional manifold (see Fig. 2), these subspaces will provide for its piecewise linear approximation. Let ω_i $(i = 1, 2, \ldots, c)$ be a class random variable and let S_i be the its principal component subspace. Then, the ith class data is generally projected best in S_i. The same subspace should be optimal for classification purpose too, but this is not necessarily true and further tests are required. In our case, in order to classify data from ω_i, the subspaces S_j $(j \neq i)$ should be explored as well. The classical PCA on the other hand approximates data using a single low-dimensional subspace (Fig. 2e) and does not use any class-specific information, for example the manifold curvature. LDA, on the contrary, uses classwise information, but results in a linear subspace (Fig. 2f). Instead of using complex linear techniques, a simple piecewise linear method may be used to efficiently reduce data dimension.

Our method is implemented through two algorithms. Using training data, the first algorithm estimates the piecewise linear subspace and comprises the training process. The second algorithm chooses an optimal subspace for feature representation and successive classification of an unknown (test) data.

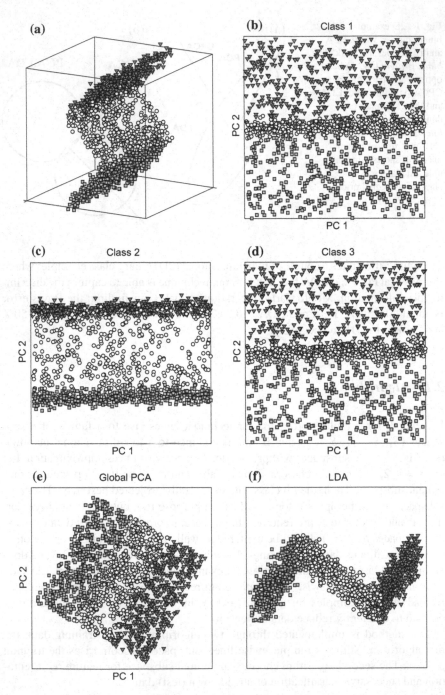

Fig. 2 **a** Data from 3 classes, labeled by □ (ω_1), ○ (ω_2) and ∇ (ω_3), is principally confined to a 2D manifold. **b** Projection of data to the 2D principal component subspace of class ω_1, **c** class ω_2 and **d** class ω_3. **e** Projection of data to the global principal component subspace. **e** Features extracted using LDA

2.1.1 Training Algorithm

Algorithm 1 (*Training*)

1. **begin** compute Σ_b, W_b (optional)
2. **initialize** $i \longleftarrow 0$
2.   **for** $i \longleftarrow i + 1$
3.   **classwise PCA:** compute W_i^{CPCA}
4.   **augment:** $E_i := \left[W_i^{CPCA} \mid W_b \right]$
5.   **discriminant:** compute T_i^{DFE} (optional)
6.   compute $F_i = E_i \, T_i^{DFE}$
7.   **until** $i = c$
8. **return** $F_1, F_2, \ldots, F_c$
9. **end**

2.1.2 Training Algorithm

Definition of Variables: Let the input be $\mathcal{X}_i$, a collection of data x belonging to class ω_i. The class mean, $\mu_i \in \mathbb{R}^n$, and covariance, $\Sigma_i \in \mathbb{R}^{n \times n}$, are defined as: $\mu_i \triangleq 1/|\mathcal{X}_i| \sum_{x \in \mathcal{X}_i} x$ and $\Sigma_i \triangleq 1/(|\mathcal{X}_i| - 1) \sum_{x \in \mathcal{X}_i} (x - \mu_i)(x - \mu_i)^T$, respectively, where $| \cdot |$ denotes the cardinality of a set and n is the dimension of the data vector. The prior state probabilities are estimated empirically from the data, i.e., $P(\omega_i) = |\mathcal{X}_i| / \sum_{j=1}^{c} |\mathcal{X}_j|$.

Firstly, the between-class scatter matrix is computed as

$$\Sigma_b = \sum_{i=1}^{c} P(\omega_i)(\mu_i - \mu)(\mu_i - \mu)^T \in \mathbb{R}^{n \times n} \tag{1}$$

where $P(\omega_i)$ is the class probability, μ_i is the sample mean of class ω_i, μ is the (overall) sample mean, defined by

$$\mu = \sum_{i=1}^{c} P(\omega_i)\mu_i \in \mathbb{R}^{n \times 1}. \tag{2}$$

The matrix $W_b \in \mathbb{R}^{n \times d}$ is computed such that its columns form an orthonormal basis for the range of Σ_b, defined as $\mathcal{R}(\Sigma_b) = \left\{ \xi \in \mathbb{R}^n : \xi = \Sigma_b \eta, \forall \eta \in \mathbb{R}^n \right\}$. It follows from (1) and (2) that Σ_b has maximum $c - 1$ linearly independent columns, therefore $d \leq c - 1$. The computation W_b is optional, and is explained later.

The algorithm then iterates over classes, and each time the principal component subspace S_i is estimated. The m_i' principal eigenvectors of the class covariance $\Sigma_i \in \mathbb{R}^{n \times n}$ forms the columns of the matrix $W_i^{CPCA} \in \mathbb{R}^{n \times m_i'}$, which represents a basis for S_i. The choice of m_i' is left to the user and several options for retaining number of

Fig. 3 Two classes marked by □ (ω_1) and ⊙ (ω_2), whose principal directions are nearly parallel

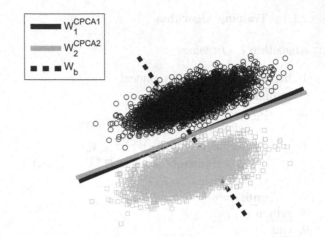

eigenvectors/eigenvalues are provided. This includes selecting number of eigenvectors based on the shape of eigenvalue spectrum, retaining eigenvalues higher than the median/mean of nonzero eigenvalues or keeping eigenvalues that capture certain percentage of the cumulative energy. In the next step, the matrix $E_i \in \mathbb{R}^{n \times (m_i' + d)}$ is formed by concatenating W_i^{CPCA} and W_b. The columns of E_i are orthonormalized through the Gram-Schmidt procedure resulting in orthogonal projections. Omission of step 2.1.1 renders the definition of E_i trivial and reduces to $E_i = W_i^{CPCA}$ (formally $d = 0$). W_b is included in the algorithm to take into account the cases where discriminatory information is present in the class means and not in the class covariances. One such example is Fig. 3, where the projections of data in the PCA subspace is highly overlapped, and the addition of W_b results in discriminatory information.

Typically, m_i' is so chosen that $m_i' + d \ll n$, and small sample size problem and curse of dimensionality are handled by projecting data to a basis defined by E_i. Following the dimensionality reduction, classical discriminant feature extraction (DFE) techniques, such as LDA, can now be applied at this stage (step 2.1.1). The use of a DFE technique, although optional, may further boost classification performance. When a *linear* DFE technique is chosen, the overall feature extraction matrix $F_i \in \mathbb{R}^{n \times m_i}$ is defined as $F_i = E_i T_i^{DFE}$ where $T_i^{DFE} \in \mathbb{R}^{(m_i' + d) \times m_i}$ is a feature extraction matrix of the linear DFE method used, and m_i is the final dimension of the feature space which can be specified by the user. If step 2.1.1 is not used, the feature extraction matrix becomes $F_i = E_i$ (formally $T_i^{DFE} = I$, where I is the identity matrix). Our algorithm always yields c feature extraction matrices, whose columns provide bases for c subspaces. Examples shown in Fig. 2b–d are obtained by applying Algorithm 1 without the optional step 2.1.1 and with $m_1' = m_2' = m_3' = 2$.

2.1.3 Optimal Subspace Algorithm

Let the high dimensional test data be $\bar{x} \in \mathbb{R}^n$. Ultimately, $\bar{x}$ should be classified in one out of c subspaces estimated using Algorithm 1. The following algorithm describes the procedure of selecting the optimal feature space.

Algorithm 2 (*Optimal Subspace*)

1. **begin** <u>**initialize**</u> $\bar{x} \longleftarrow$ test vector, $F_1, F_2, \ldots, F_c \longleftarrow$ feature extraction matrices, $i \longleftarrow 0$
2.   <u>**for**</u> $i \longleftarrow i+1$
3.   project $\bar{x}^i = F_i^T \bar{x}$
4.   <u>**initialize**</u> $j \longleftarrow 0$
5.   <u>**for**</u> $j \longleftarrow j+1$
6.   **Bayes rule:** compute $P(\omega_j^i \mid \bar{x}^i) = \dfrac{f(\bar{x}^i \mid \omega_j^i)P(\omega_j)}{f(\bar{x}^i)}$
7.   <u>**until**</u> $j = c$
8.   maximize over classes $J(i) = \arg \max_{1 \le j \le c} P(\omega_j^i \mid \bar{x}^i)$
9.   <u>**until**</u> $i = c$
10.   maximize over subspaces $I = \arg \max_{1 \le i \le c} P(\omega_{J(i)}^i \mid \bar{x}^i)$
11. <u>**return**</u> I, J(I)
12. **end**

Algorithm 2 estimates the optimal subspace. The outer loop goes over all the c subspaces and in each such subspace, spanned by the columns of $F_i \in \mathbb{R}^{n \times m_i}$, a projection, $\bar{x}^i \in \mathbb{R}^{m_i}$, of $\bar{x}$ is found. The algorithm then loops over classes, and the posterior probability of class ω_j given the projection of $\bar{x}$ is found via the Bayes rule for each subspace. This step results in projecting (training) data from each class onto the subspace at hand, and estimating the class-conditional probability density function (PDF), denoted by $f(. \mid \omega_j^i)$. We approximate these PDFs with Gaussian densities, i.e., $f(. \mid \omega_j^i) \sim \mathcal{N}(F_i^T \mu_j, F_i^T \Sigma_j F_i)$. This approach is consistent with Algorithm 1, which also used second-order sample statistics, namely μ_j and Σ_j. Hence, this approach is best suited for Gaussian classes, although it should also give good results in cases when the discriminatory information between classes lie in the first two statistical moments. When the data violates the Gaussian assumption, different approaches may be preferable, such as using the mixture density approach [16], or the kernel density approach [46]. However, these methods result in a higher computational cost compared to the Gaussian assumption approach. Irrespective of the PDF estimation technique, the mathematical formalism of step 2.1.3 and subsequent steps, remains the same.

The algorithm then computes the maximum a posteriori (MAP) in each subspace. In other words, the most likely class is found given the projection of $\bar{x}$, and the class labels of these locally optimal classes are denoted by $J(i)$. Each subspace may end up with a different class having the maximum posterior probability. The final class membership is then found by using the MAP rule to the local winners (step 2.1.3), and the winning subspace is marked by I. The MAP rule (step 2.1.3) holds for different dimensional subspaces too since the posterior probabilities are normalized. The feature extraction matrix for $\bar{x}$ is given by the basis F_I of the winning subspace and test feature, $F_I^T \bar{x}$, to be used for final classification is the projection of the test data $\bar{x}$ onto the chosen subspace. After the feature extraction process is completed, the test features can be classified by applying any chosen pattern classifiers.

2.1.4 Algorithm Summary

The proposed method is possibly an improvement over single-subspace techniques, especially where high-dimensional data lies in a low-dimensional manifold which cannot be approximated by a linear subspace. CPCA-based method is expected to perform better compared to other techniques on such hard classification problems. Moreover, LDA-based methods suffer from the feature subspace dimension constraint, $m_i <= c - 1$, whereas CPCA-based methods are more flexible in the choice of m_i, thereby giving a better performance. Comparisons of CPCA and other competing techniques using different datasets are given in [11].

While the number of principal components retained in each class may differ over classes, the final feature dimension, m_i, is kept constant for the sake of simplicity and comparison with other methods. The choice of m_i can also differ for each class under the CPCA framework without changing the mathematical formalism. Using different m_i, however, might lead to a combinatorial problem and is hence not pursued.

There are a couple of drawbacks of the proposed method. The time complexity of CPCA is $O(c)$ where c is the number of possible subspaces, and hence the method is not computationally efficient for large number of classes. However, within a single subspace, the feature extraction scheme is linear and reduces to a simple eigen decomposition, which can be implemented efficiently using matrix manipulations. The CPU times for different datasets using the proposed method are reported in [11] and demonstrate its computational efficiency. While the technique scales unfavorably with number of classes, c, its performance is comparable to the other methods. One other weakness of the proposed method is that by keeping the m_i' eigenvectors of the class covariance, Σ_i, only the regularity of the class covariance, Σ_i, is guaranteed. However, there is no assurance that the other class covariance matrices, $\Sigma_j, (j \neq i)$, will remain nonsingular in the subspace S_i. A potential solution is to shrink S_i (by reducing m_i') until all the class covariances become regular. Clearly, if $S_i \subset N(\Sigma_j)(j \neq i)$, where $N(\Sigma_j)$ is the null space of Σ_j, no choice of m_i' would render Σ_j regular in S_i, and the method may fail. However, based on the analyzed neural datasets and datasets in [11], this scenario seems highly unlikely. Interested readers can refer to [11] for detailed analysis of the proposed method.

3 Experiments with Big Datasets in Brain–Computer Interface

The previous section discuss different feature extraction and classification techniques that can be used on large-scale data for classification purpose. Now, we shift our focus to the application of our methods in the context of BCI.

3.1 BCI System

Successful BCI experiments have been reported using both invasive [21, 33, 41, 58, 62, 71] and noninvasive brain recording technologies [4, 28, 47, 48, 72, 74]. These results hold promise for people with severe motor disabilities, such as those suffering from amyotropic lateral sclerosis (ALS), brainstem stroke, or other neuromuscular diseases. While most of these patients have little, or no muscular control, typically their cognitive functions remain unaffected by the paralysis. The goal of BCI is to convert these thoughts into actions. While promising experimental results have been obtained [4, 21, 33, 41, 48, 58, 62, 71, 73, 74], a fully operational practical BCI remains a challenging task.

In this section we describe a standard memory-reach experimental setup and demonstrate the efficacy of CPCA in decoding neural signals.

Given a set of brain data, the *simplest* fully operational BCI system must be able to answer with certainty at least two questions:

1. Is the subject's intention to move or not?
2. If the movement is intended, should the subject move to the left or right?

For simplicity, we will refer to these two questions as the *when* and *where* questions, respectively. Consequently, the BCI system can be modeled as a finite state machine (FSM) with *Idle*, i.e., "do-not-move" state as well as *Move* state (see Fig. 4). Transitions between the states are initiated with each newly acquired block of brain data

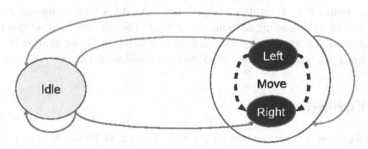

Fig. 4 FSM diagram of a minimal practical BCI system. Allowed state transitions are denoted by *arrows*

and are executed in real time. For example, based on the most recent data analysis, the system in the *Idle* state can retain the present state (self-pointing arrow) or make a transition to the *Move* state. It should be noted here that in case of paralyzed patients, the *Move* state should be replaced with *Intent to Move* state. Also note that the *Move* state consists of two mutually exclusive substates: *Left* and *Right*. Clearly, the *Left* and *Right* substates can be replaced with any other dichotomy, such as "Up/Down," "Yes/No," "True/False," and similar. Finally, we note that a BCI system can have more than two states (see [61] for example) and/or substates, although the presence of additional states will necessarily increase the number of parameters and may result in higher error rates.

Here, we apply our proposed technique to answer the *where* questions under a standard memory-reach experimental setup. In the following sections, we will discuss the experiment and results using our method to decode the direction i.e., left or right. Our techniques can serve as a general analytical tool for any large-scale data like images or videos. We have also applied our methods for various other applications including face recognition and digit recognition, cancer detection, and many others (see [11] for details).

3.2 Direction Estimation

Decoding directional activities using neural signals has been one of the main focus areas of BCI research [10, 41, 47, 52, 54, 61]. BCIs equipped with efficient directional tuning capabilities can assist paralyzed patients in performing simple tasks using output devices like robotic arm, thus restoring some of the impaired motor functions. Furthermore, techniques used for successful estimation of direction can serve as analytical tools to understand and analyze spatial processing in the brain and provide useful insights related to brain research. In many BCI approaches [12, 41, 52, 61], subjects are instructed to carry out planned-reach movements while their neural activities are recorded. Decoding the direction of such planned-reach movements essentially answers the "where" question (Sect. 3.1). Here we use ECoG recordings of humans, using standard memory-reach setup, to decode the two possible directions (left or right) of target appearing on screen using our proposed approaches. In the following sections, we describe the experimental setup for direction estimation, followed by results using our methods. Finally we analyze the results obtained and discuss the key observations with concluding remarks.

3.2.1 Experimental Setup

The performance of our method was tested on a set of ECoG data of a severely epileptic patient, adopted from Rizzuto et al. [52], and an EEG dataset of normal subjects recorded in our lab. The ECoG signals were recorded from the human brain during a standard memory-reach task, consisting of 4 periods: *fixation, target, delay,*

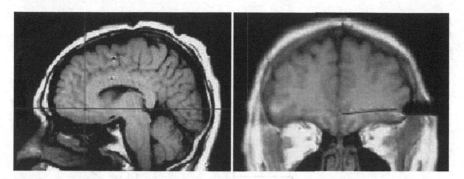

Fig. 5 *Left* Sagital MRI image showing electrode placement in subject. Electrodes were implanted orthogonally in a lateral to medial trajectory perpendicular to the intersection of the *thin red lines*. *Right* Coronal view of depth electrode placement in the same subject. The electrode trajectory is shown in *red*

and *reach*. The experiment consisted of the subject making directed arm movements to different locations (left or right) on a touchscreen interface (see Figs. 5 and 6). In the beginning of each trial, a fixation stimulus is presented in the center of the touchscreen and the participant initiates the trial by placing his right hand on the stimulus. After a short *fixation* period, a *target* is flashed on the screen, followed by a *memory* period. The fixation is extinguished after the *memory* period, which acts as a cue for the participants to *reach* to the memorized location indicated previously by the *target*. The *fixation*, *target* and *memory* periods are varied uniformly between 1 and 1.3 s. The subject had several electrodes implanted into each of the following target brain areas: orbital frontal (OF) cortex, amygdala (A), hippocampus (H), anterior cingulate (AC) cortex, supplementary motor (SM) cortex, and parietal (P) cortex. The total number of electrodes implanted in both hemispheres was 91 and 162 trials were recorded for each period. The signals were amplified, sampled at 200 Hz, and band-pass filtered. The goal of our study is to predict the label of the trial (left vs. right) based on 1 s of data during the four periods. Note that data is a vector in 18200-dimensional space ($n = 91 \times 200$).

For the EEG experiments, a similar setup adapted from [52] was used, with two periods: *fixation* and *target* (see Fig. 7). The EEG signals were acquired using an EEG cap (Electro-Cap International, Eaton, OH) with 6 electrodes placed in occipital and parietal regions following the international 10–20 system, and the signals were amplified, band-pass filtered, and sampled at 200 Hz (Biopac Systems, Goleta, CA). The number of trials (left + right) was $N = 140$ per session, and there were 3 such sessions. Our goal is to predict the label of the trial (left vs. right) based on 1 s of data.

Fig. 6 Experimental setup for ECoG data for decoding directions

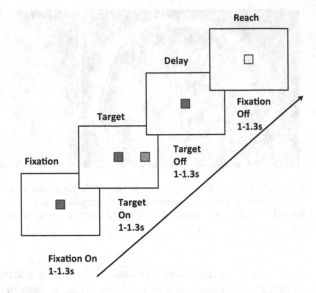

Fig. 7 Experimental setup for EEG data for decoding directions

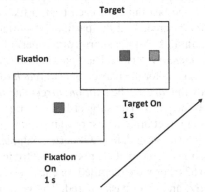

3.2.2 Results

We are decoding the location of target (left/right) from the neural data and the output of our analyses could potentially be used to decode a person's motor intentions in real time. The ECoG and EEG data were classified using CPCA. The data was also decoded using DLDA [76] for comparison purpose. Quadratic classifier was used for all decoding results. Principal components with eigenvalues exceeding the mean eigenvalue of the covariance matrix for each class were retained in CPCA. Leave-one-out cross validation was used for reporting the performance. Using validatory data to find the final reduced dimension results in some degree of overfitting. Thus the reported classification rates are not generalizable. The performance reported are feasible and suffice for comparison purpose. Leave-one-out CV produces nearly unbiased error estimates [32] as compared to k fold CV and random

Table 1 The performances (%) of CPCA and DLDA for ECoG data, during *target*, *delay* and *reach* period. (Top) Unconstrained, (bottom) constrained data

Period	n	Time	Brain area	DLDA	CPCA
Target	18200	All	All	70.37	*79.01*
Delay	18200	All	All	58.02	*64.82*
Reach	18200	All	All	66.05	*67.28*
Delay	9600	All	SM, P, OF	65.43	*74.69*

Table 2 The performances (%) of CPCA, DLDA and threshold-based DLDA for EEG data, during *target* period

Session	n	Time	DLDA	CPCA
1	1200	All	52.14	**75.71**
1	180	100:250	60.71	**90.00**
2	1200	All	50.71	**62.14**
2	180	100:250	55.00	**82.14**
3	1200	All	57.24	**75.35**
3	180	100:250	60.14	**90.58**

split methods [16]. The performance using leave-one-out CV can be reproduced and multiple runs is not required for computing good error estimates. Hence leave-one-out CV is best suited for comparing various feature extraction methods.

The estimated classification rates for the ECoG and EEG datasets are reported in Tables 1 and 2, respectively. Additional results with more subjects is illustrated in [12, 13]. We make the following observations based on these results. Our CPCA-based classification method performs well and produces the best classification in all cases. CPCA method sometimes produces an improvement of around 50 % when compared with DLDA-based method, using both unconstrained and constrained data. Typically the first 100–150 ms of the target period can be attributed to visual processing delay [64] and can be discarded. Similarly based on brain areas of interest, subsets of electrodes can be chosen.

LDA-based methods perform poorly giving almost chance performance, especially for EEG data, whereas our method results in better classification. Only 1-D features were used for CPCA to make it comparable to DLDA since LDA-based methods can have at most $(c - 1)$-D subspaces ($c = 2$ here). CPCA using larger feature dimension(e.g., $m = 2, 3$) produced marginally better results. LDA-based techniques fare poorly on noisy neural data and gives better results when the data is comparatively less noisy as in face recognition, object recognition tasks.

3.2.3 Discussion

All analysis reported in the Tables 1 and 2 process the data without assuming space-time separability and analyzing the filters produced by these methods, we are able to comment on the important brain areas and time instants, which carries discriminatory information and helps in identifying the two target directions, *left, right*. The spatio-temporal filter has the same dimension as the original data having rows equal to the number of electrode channels and columns denoting time. We analyze the filters produced by CPCA and discuss some key aspects of these filters.

Figure 8 shows the the spatio-temporal filters, obtained using CPCA, corresponding to 1 s of ECoG data during the target period. Note here that we use a piecewise linear method and unlike DLDA which generates a single subspace, our technique

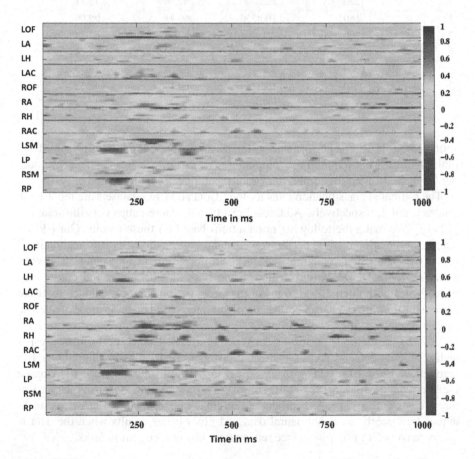

Fig. 8 Spatio-temporal filters for ECoG data for target period using CPCA. *Top* figure illustrates Subspace 1 and *bottom* figure shows subspace 2

here generates 2 subspaces (one for each class) which acts as a nonlinear manifold approximating the high dimensional brain data in the original space.

The prefix L or R denotes the left and the right hemisphere. Our method produces a spatio-temporal filter and there is no need to separately process space and time. Most informative brain area and time epoch for decoding the location of the target are given by large positive (dark red) and large negative (dark blue) coefficients. Most of the information is present ~150–200 ms poststimulus (onset of the target is at 0), which is consistent with the visual information processing delay [64].

From the distribution of important coefficients in Fig. 8, discriminatory information appears to be localized in time and space, while the noise is distributed, as shown by large areas having near-zero coefficients (light blue, green and yellow). In Fig. 8, some spatio-temporal areas within the stimulus onset appear 100 ms, which is in contradiction to the latency of visual information processing (~150 ms [64]). Discriminatory information is not expected so early on in the trial, and these coefficients mostly represent noise artifacts, such as biological noise, recording hardware noise or ambient noise.

From Fig. 8, it follows that the Parietal and Hippocampal regions play in important role in distinguishing the two target locations, left and right. The findings are consistent with the cortical studies as Hippocampus and Parietal regions are known to be associated with the spatial location [27]. Orbitofrontal region (OF), associated with decision making, and Supplementary Motor Area (SMA), associated with planning movement, also shows up as important regions. Since the *target* state is followed by the *delay* state, where one would expect OF and SMA to exhibit prominent contribution, the finding is not surprising.

It can be concluded based on the results that our proposed method efficiently decodes large-scale neural data, suffering from small sample size problem. We have successfully applied our technique to decode direction of planned-reach movements in human using ECoG and EEG recordings. Space-time separability assumption is not required in our method and it can be used as an analysis tool to explore the any generic big data.

With the advent of large-scale recording technologies and collaborative data collection efforts in neuroscience, efficient dimensionality reduction techniques will be crucial to the big data analysis. Only time will tell the effect big data has on our understanding of neural processes and the insights it provide. We hope that we can solve the challenges related to big data analysis and it provides us with thought-provoking results in the coming days.

References

1. Focus on big data. Nat Neurosci 17(11):1429 http://www.nature.com/neuro/journal/v17/n11/full/nn.3856.html
2. Ang KK, Chin ZY, Zhang H, Guan C (2008) Filter bank common spatial pattern (fbcsp) in brain-computer interface. In: IEEE International joint conference on neural networks, 2008. IJCNN 2008 (IEEE World congress on computational intelligence). IEEE, pp 2390–2397

3. Belhumeur P, Hespanha J, Kriegman D (1997) Eigenfaces vs. fisherfaces: recognition using class specific linear projection. IEEE Trans Pattern Anal 19(7):711–720
4. Birbaumer N, Ghanayim N, Hinterberger T, Iversen I, Kotchoubey B, Kbler A, Perelmouter J, Taub E, Flor H (1999) A spelling device for the paralysed. Nature 398(6725):297–298
5. Blankertz B, Tomioka R, Lemm S, Kawanabe M, Muller K-R (2008) Optimizing spatial filters for robust eeg single-trial analysis. IEEE Signal Process Mag 25(1):41–56
6. Chen L-F, Liao H-YM, Ko M-T, Lin J-C, Yu G-J (2000) A new lda-based face recognition system which can solve the small sample size problem. Pattern Recogn 33(10):1713–1726
7. Cover TM, Campenhout JMV (1977) On the possible ordering in the measurement selection problem. IEEE Trans Syst Man Cybern 7:657–661
8. Daly JJ, Cheng R, Rogers J, Litinas K, Hrovat K, Dohring M (2009) Feasibility of a new application of noninvasive brain computer interface (bci): a case study of training for recovery of volitional motor control after stroke. J Neurol Phys Ther 33(4):203–211
9. Daniels MJ, Kass RE (2001) Shrinkage estimators for covariance matrices. Biometrics 57:1173–1184
10. Das K, Meyer J, Nenadic Z (2006) Analysis of large-scale brain data for brain-computer interfaces. In: Proceedings of the 28th Annual international conference of the IEEE engineering in medicine and biology society, pp 5731–5734
11. Das K, Nenadic Z (2009) An efficient discriminant-based solution for small sample size problem. Pattern Recogn 42(5):857–866
12. Das K, Osechinskiy S, Nenadic Z (2007) A classwise pca-based recognition of neural data for brain-computer interfaces. In: Proceedings of the 29th Annual international conference of the IEEE engineering in medicine and biology society, pp 6519–6522
13. Das K, Rizzuto D, Nenadic Z (2009) Estimating mental states for brain-computer interface. IEEE Trans Biomed Eng 56(8):2114–2122
14. Do AH, Wang PT, King CE, Chun SN, Nenadic Z (2013) Brain-computer interface controlled robotic gait orthosis. J Neuro Eng Rehabil 10(111)
15. Do AH, Wang PT, King CE, Schombs A, Cramer SC, Nenadic Z (2012) Brain-computer interface controlled functional electrical stimulation device for foot drop due to stroke. Conf Proc IEEE Eng Med Biol Soc 6414–6417:2012
16. Duda RO, Hart PE, Stork DG(2001) Pattern classification. Wiley-Interscience
17. Enke H, Partl A, Reinefeld A, Schintke F (2012) Handling big data in astronomy and astrophysics: rich structured queries on replicated cloud data with xtreemfs. Datenbank-Spektrum 12(3):173–181
18. Fisher RA (1936) The use of multiple measurements in taxonomic problems. Ann Eugenics 7:179–188
19. Fukunaga K (1990) Inroduction to statistical pattern recognition, 2nd edn. Academic Press
20. He X, Yan S, Hu Y, Niyogi P (2005) Face recognition using laplacian faces. IEEE Trans Pattern Anal 27(3):328–340
21. Hochberg L, Serruya M, Friehs G, Mukand J, Saleh M, Caplan A, Branner A, Chen D, Penn RD, Donoghue J (2006) Neuronal ensemble control of prosthetic devices by a human with tetraplegia. Nature 442(7099):164–171
22. Hoffbeck JP, Landgrebe DA (1996) Covariance matrix estimation and classification with limited training data. IEEE Trans Pattern Anal 18(7):763–767
23. Huang R, Liu Q, Lu H, Ma S (2002) Solving the small sample size problem of lda. In: ICPR '02: Proceedings of the 16th International conference on pattern recognition (ICPR'02), vol 3. IEEE Computer Society, p 30029
24. Huber PJ (1985) Projection pursuit. Ann Stat 13(2):435–475
25. Insel TR, Landis SC, Collins FS (2013) The nih brain initiative. Science 340(6133):687–688
26. Kawanabe M, Samek W, Müller K-R, Vidaurre C (2014) Robust common spatial filters with a maxmin approach. Neural Comput 26(2):349–376
27. Kesner RP, Farnsworth G, Kametani H (1991) Role of parietal cortex and hippocampus in representing spatial information. Cereb Cortex 1(5):367–373

28. King CE, Dave KR, Wang PT, Mizuta M, Reinkensmeyer DJ, Do AH, Moromugi S, Nenadic Z (2014) Performance assessment of a brain-computer interface driven hand orthosis. Annals of Biomed Eng 42(10):2095–2105

29. King CE, Wang PT, McCrimmon CM, Chou CCY, Do AH, Nenadic Z (2014) Brain-computer interface driven functional electrical stimulation system for overground walking in spinal cord injury participant. In: Proceedings of the 36th Annual international conference IEEE engineering and medical biological society, pp 1238–1242

30. Kirby M, Sirovich L (1990) Application of the karhunen-loeve procedure for the characterization of human faces. IEEE Trans Pattern Anal 12:103–108

31. Kittler J (1978) Feature set search algorithms. In: Chen CH (ed) Pattern recognition and signal processing. Sijthoff and Noordhoff, Alphen aan den Rijn, The Netherlands, pp 41–60

32. Kohavi R (1995) A study of cross-validation and bootstrap for accuracy estimation and model selection. In: Joint commission international, pp 1137–1145

33. Leuthardt E, Schalk G, Wolpaw J, Ojemann J, Moran D (2004) A brain-computer interface using electrocorticographic signals in humans. J Neural Eng 1(2):63–71

34. Lewis S, Csordas A, Killcoyne S, Hermjakob H, Hoopmann MR, Moritz RL, Deutsch EW, Boyle J (2012) Hydra: a scalable proteomic search engine which utilizes the hadoop distributed computing framework. BMC Bioinform 13(1):324

35. Loog M, Duin R (2004) Linear dimensionality reduction via a heteroscedastic extension of lda: the chernoff criterion. IEEE Trans Pattern Anal 26:732–739

36. Markram H (2012) The human brain project. Sci Am 306(6):50–55

37. Mathé C, Sagot M-F, Schiex T, Rouze P (2002) Current methods of gene prediction, their strengths and weaknesses. Nucleic Acids Res 30(19):4103–4117

38. McCrimmon CM, King CE, Wang PT, Cramer SC, Nenadic Z, Do AH (2014) Brain-controlled functional electrical stimulation for lower-limb motor recovery in stroke survivors. Conf Proc IEEE Eng Med Biol Soc 2014:1247–1250

39. McFarland DJ, McCane LM, David SV, Wolpaw JR (1997) Spatial filter selection for EEG-based communication. Electroeng Clin Neuro 103(3):386–394

40. Müller-Gerking J, Pfurtscheller G, Flyvbjerg H (1999) Designing optimal spatial filters for single-trial eeg classification in a movement task. Clin Neurophysiol 110(5):787–798

41. Musallam S, Corneil BD, Greger B, Scherberger H, Andersen RA (2004) Cognitive control signals for neural prosthetics. Science 305(5681):258–262

42. Nenadic Z (2007) Information discriminant analysis: Feature extraction with an information-theoretic objective. IEEE Trans Pattern Anal 29(8):1394–1407

43. Nenadic Z, Rizzuto D, Andersen R, Burdick J (2007) Advances in cognitive neural prosthesis: recognition of neural data with an information-theoretic objective. In: Dornhege G, Millan J, Hinterberger T, McFarland D, Muller KR (eds) Toward brain computer interfacing, Chap 11. The MIT Press, pp 175–190

44. Nilsson T, Mann M, Aebersold R, Yates JR III, Bairoch A, Bergeron JJ (2010) Mass spectrometry in high-throughput proteomics: ready for the big time. Nat Methods 7(9):681–685

45. ODriscoll A, Daugelaite J, Sleator RD (2013) Big data, hadoop and cloud computing in genomics. J Biomed Inf 46(5):774–781

46. Parzen E (1962) On the estimation of a probability density function and mode. Ann Math Stat 33:1065–1076

47. Pfurtscheller G, Neuper C, Flotzinger D, Pregenzer M (1997) EEG-based discrimination between imagination of right and left hand movement. Electroeng Clin Neuro 103(6):642–651

48. Pfurtscheller G, Neuper C, Muller GR, Obermaier B, Krausz G, Schlogl A, Scherer R, Graimann B, Keinrath C, Skliris D, Wortz GSM, Schrank C (2003) Graz-BCI: state of the art and clinical applications. IEEE Trans Neural Syst Rehabil 11(2):177–180

49. Ramos-Murguialday A, Broetz D, Rea M, Ler L, Yilmaz O, Brasil FL, Liberati G, Curado MR, Garcia-Cossio E, Vyziotis A, Cho W, Agostini M, Soares E, Soekadar S, Caria A, Cohen LG, Birbaumer N (2013) Brain-machine interface in chronic stroke rehabilitation: a controlled study. Ann Neurol 74(1):100–108

50. Ramoser H, Muller-Gerking J, Pfurtscheller G (2000) Optimal spatial filtering of single trial eeg during imagined hand movement. IEEE Trans Rehabil Eng 8(4):441–446
51. Rao CR (1948) The utilization of multiple measurements in problems of biological classification. J R Stat Soc B Methods 10(2):159–203
52. Rizzuto D, Mamelak A, Sutherling W, Fineman I, Andersen R (2005) Spatial selectivity in human ventrolateral prefrontal cortex. Nat Neurosci 8:415–417
53. Roweis ST, Saul LK (2000) Nonlinear dimensionality reduction by locally linear embedding. Science 290:2323–2326
54. Santhanam G, Ryu SI, Yu1 BM, Afshar A, Shenoy KV (2006) A high-performance brain-computer interface. Nature 442:195–198
55. Saon G, Padmanabhan M (2000) Minimum Bayes error feature selection for continuous speech recognition. In: NIPS, pp 800–806
56. Schäfer J, Strimmer K (2005) A shrinkage approach to large-scale covariance matrix estimation and implications for functional genomics. Stat Appl Genet Mol Biol 4(1):Article 32
57. Schölkopf B, Smola A, Müller KR (1997) Kernel principal component analysis. In: Artificial neural networks ICANN'97. Springer, pp 583–588
58. Serruya MD, Hatsopoulos NG, Paninski L, Fellows MR, Donoghue JP (2002) Instant neural control of a movement signal. Nature 416(6877):141–142
59. Shenoy P, Miller K, Ojemann J, Rao R (2007) Finger movement classification for an electrocorticographic bci. In: 3rd International IEEE/EMBS conference on neural engineering, 2007. CNE '07, pp 192–195, 2–5 May 2007
60. Shenoy P, Miller K, Ojemann J, Rao R (2008) Generalized features for electrocorticographic bcis. IEEE Trans Biomed Eng 55(1):273–280
61. Shenoy P, Rao R (2004) Dynamic bayes networks for brain-computer interfacing. In: Advances in neural information processing systems, pp 1265–1272
62. Taylor D, Tillery S, Schwartz A (2002) Direct cortical control of 3D neuroprosthetic devices. Science 296(5574):1829–1832
63. Thomaz CE, Gillies DF, Feitosa RQ (2001) Small sample problem in bayes plug-in classifier for image recognition. In: International conference on image and vision computing, New Zealand, pp 295–300
64. Thorpe S, Fize D, Marlot C (1995) Speed of processing in the human visual system. Nature 381(6582):520–522
65. Torkkola K (2002) Discriminative features for document classification. In: Proceedings 16th international conference on pattern recognition, vol 1, pp 472–475
66. Townsend G, Graimann B, Pfurtscheller G (2006) A comparison of common spatial patterns with complex band power features in a four-class bci experiment. IEEE Trans Biomed Eng 53(4):642–651
67. Turk M, Pentland A (1991) Eigenfaces for recognition. J Cognit Neurosci 3(1):71–86
68. Wang W, Collinger JL, Degenhart AD, Tyler-Kabara EC, Schwartz AB, Moran DW, Weber DJ, Wodlinger B, Vinjamuri RK, Ashmore RC et al (2013) An electrocorticographic brain interface in an individual with tetraplegia. PloS ONE 8(2):e55344
69. Wang X, Tang X (2004) Dual-space linear discriminant analysis for face recognition. In: 2004 IEEE Computer society conference on computer vision and pattern recognition (CVPR'04), vol 02, pp 564–569, 2004
70. Wang Y, Gao S, Gao X (2006) Common spatial pattern method for channel selelction in motor imagery based brain-computer interface. In: 27th Annual international conference of the engineering in medicine and biology society, 2005. IEEE-EMBS 2005. IEEE, pp 5392–5395
71. Wessberg J, Stambaugh C, Kralik J, Beck P, Laubach M, Chapin J, Kim J, Biggs S, Srinivasan MA, Nicolelis MA (2000) Real-time prediction of hand trajectory by ensembles of cortical neurons in primates. Nature 408(6810):361–365
72. Wolpaw J, Birbaumer N, McFarland D, Pfurtscheller G, Vaughan T (2002) Brain-computer interfaces for communication and control. Clin Neurophysiol 6(113):767–791
73. Wolpaw J, Wolpaw EW (2011) Brain-computer interfaces: principles and practice. Oxford University Press

74. Wolpaw JR, McFarland DJ (2004) Control of a two-dimensional movement signal by a noninvasive brain-computer interface in humans. Proc Natl Acad Sci USA 101(51):17849–17854
75. Yang J, Frangi AF, Yang J-Y, Zhang D, Jin Z (2005) Kpca plus lda: a complete kernel fisher discriminant framework for feature extraction and recognition. IEEE Trans Pattern Anal Mach Intell 27(2):230–244
76. Yu H, Yang H (2001) A direct lda algorithm for high-dimensional data with application to face recognition. Pattern Recogn Lett 34(10):2067–2070
77. Zhang S, Sim T (2007) Discriminant subspace analysis: a fukunaga-koontz approach. IEEE Trans Pattern Anal 29(10):1732–1745
78. Zhu M, Martinez AM (2006) Subclass discriminant analysis. IEEE Trans Pattern Anal 28(8):1274–1286

Big Data and Cancer Research

Binay Panda

1 Introduction

The advent of high-throughput technology has revolutionized biological sciences in the last two decades enabling experiments on the whole genome scale. Data from such large-scale experiments are interpreted at system's level to understand the interplay among genome, transcriptome, epigenome, proteome, metabolome, and regulome. This has enhanced our ability to study disease systems, and the interplay between molecular data with clinical and epidemiological data, with habits, diet, and environment. A disproportionate amount of data has been generated in the last 5 years on disease genomes, especially using tumor tissues from different subsites, using high-throughput sequencing (HTS) instruments. Before elaborating the use of HTS technology in generating cancer-related data, it is important to describe briefly the history of DNA sequencing and the revolution of second and third generation of DNA sequencers that resulted in much of today's data deluge.

2 Sequencing Revolution

The history of DNA sequencing goes back to the late 1970s when Maxam and Gilbert [1] and Sanger, Nicklen and Coulson [2] independently showed that a stretch of DNA can be sequenced either by using chemical modification method or by chain termination method using di-deoxy nucleotides, respectively. Maxam and Gilbert's method of DNA sequencing did not gain popularity due to the usage of toxic chemicals and the di-deoxy chain termination method proposed by Professor

B. Panda (✉)
Ganit Labs, Bio-IT Centre, Institute of Bioinformatics and Applied Biotechnology,
Biotech Park, Electronic City Phase I, Bangalore 560100, India
e-mail: binay@ganitlabs.in

© Springer India 2016
S. Pyne et al. (eds.), *Big Data Analytics*, DOI 10.1007/978-81-322-3628-3_14

259

Fred Sanger became the de facto standard and method of choice for researchers working in the field of DNA sequencing. Many of the present-day high-throughput next-generation sequencing methods (described later) use the same principle of sequencing-by-synthesis originally proposed by Sanger. The pace, ease, and automation of the process have since grown further with the advent of PCR and other incremental, yet significant, discoveries including introduction of error-free, high fidelity enzymes, use of modified nucleotides, and better optical detection devices. It is essentially the same technology, first proposed and used by Fred Sanger [2], with modifications that led to the completion of the first draft of the Human Genome Project [3, 4] that ushered in a new era of DNA sequencing.

The idea behind some of the first generation high-throughput sequencing (HTS) assays was to take a known chemistry (predominantly the Sanger's sequencing-by-synthesis chemistry) and parallelize the assay to read hundreds of millions of growing chains of DNA rather than tens or hundreds as done with capillary Sanger sequencing. The processes for HTS comprise mainly of four distinct steps, template preparation, sequencing, image capture, and data analysis (Fig. 1). Different HTS platforms use different template preparation methods, chemistry to sequence DNA, and imaging technology that result in differences in throughput, accuracy, and running costs among platforms. As most imaging

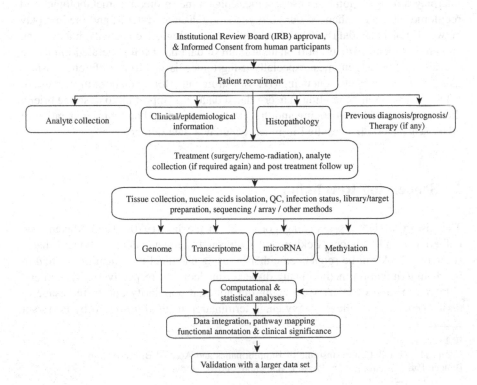

Fig. 1 Steps involved in HTS assays involving cancer patient samples and variant discovery, validation and interpretation

systems are not designed to detect single fluorescent events, clonal amplification of templates prior to imaging is incorporated as a part of template preparation before optical reading of the signal. In some cases, as in the case of single molecule sequencing, templates are not amplified but read directly to give base-level information. Some platforms are better suited than others for certain types of biological applications [5]. For discovery of actionable variants in tumors, accuracy is more important over all other parameters. Therefore, some HTS platforms are better suited to study tumor genomes over others. However, as the cost per base goes down, accuracy is increasingly achieved by higher coverage, thereby compensating errors with higher number of overlapping reads. The first publications on the human genome resequencing using the HTS system appeared in 2008 using pyrosequencing [6] and sequencing-by-synthesis using reversible terminator chemistry [7]. Since that time, the field that has gained the most amount of information using HTS platforms is cancer science. The discovery of novel DNA sequence variants in multiple cancer types using HTS platforms along with the advances in analytical methods has enabled us with the tools that have the potential to change the way cancer is currently diagnosed, treated, and managed.

3 Primary Data Generation in Cancer Studies

Various steps involved in a typical high-throughput experiment involving cancer tissue are depicted in Fig. 1. Briefly, when the patient is admitted in the hospital, clinical, epidemiological and information on habits and previous diagnosis, and treatment (if any) is recorded. Any study involving human subjects must be preapproved by an institutional review/ethics board with informed consent from all participants. Following this, analytes, full history of patients, including information on habits, and previous diagnosis/treatment (if any) are collected. Then the patients undergo treatment (surgery/chemoradiation) and the tumor tissue is collected and stored properly till further use. Once the tumor/adjacent normal/blood is collected, nucleic acids are isolated, checked for quality, and used in library/target preparation for HTS or microarray experiments. Once the raw data is collected, the data is analyzed by computational and statistical means before being integrated with clinical and epidemiological features to come up with a set of biomarkers, which is then validated in a larger cohort of patients.

4 High-Throughput Data

HTS platforms generate terabytes of data per instrument per run per week. For example, the Illumina HiSeq 4000 can generate nearly 3 terabytes of data per run in 7 days (or >400 Gb of data per day). This pose challenges for data storage, analysis, sharing, interpreting, and archiving.

Although there are many different HTS instruments in the market, the bulk of the cancer data so far have been generated using the Illumina's sequencing-by-synthesis chemistry. Therefore, a detailed description is provided on the data size, types, and complexity involved in cancer data generated by the Illumina instruments. Below is a description of different data types, usually produced during the course of a cancer high-throughput discovery study.

Despite the fact that the process of high-throughput data generation using Illumina sequencing instruments has become streamlined, never-the-less, there are inherent limitations on the quality of data generated. Some of the limitations are high degree of errors in sequencing reads (making some clinical test providers sequence up to $1000\times$ coverage or more per nucleotide to attain the requisite accuracy), shorter sequencing reads (HiSeq series of instruments do not produce data with longer than 150 nt read length), the assay not interrogating the low-complexity regions of the genome, and higher per sample cost (to gain the requisite accuracy, one needs to spend thousands of dollars per sample even for a small gene panel test). Details on different data types generated by Illumina HiSeq instrument, their approximate sizes and file type descriptions are provided in Table 1.

5 Primary Data Analysis

The cancer data analysis schema is represented in Fig. 2. First, the raw image data from the sequencing instruments are converted into *fastq* format, which is considered as the primary data files for all subsequent analysis. Before analyzing the data, the quality of the *fastq* files is checked by using tools like FastQC (http://www.bioinformatics.babraham.ac.uk/projects/fastqc/), or with *in-house* scripts to reduce sequencing quality-related bias in subsequent analysis steps. Next, sequencing reads are aligned against a reference sequence. Broadly, the alignment tools fall into two major classes, depending on which of the indexing algorithm it uses: (hash table-based or Burrows Wheeler transformation (BWT)-based). Some of the commonly used alignment tools that use hash table-based approach are Bfast [8], Ssaha [9], Smalt [10], Stampy [11] and Novoalign [12] and the ones that are BWT-based are Bowtie [13], Bowtie2 [14], and BWA [15]. BWA is the most widely used aligner by the research community. Lately, many of the alignment programs are made parallel to gain speed [16–19]. Most aligners report the results of the alignment in the form of Sequence Alignment/Map (SAM, and its binary form the BAM) format [20] that stores different flags associated with each read aligned. Before processing the aligned files for calling single (SNVs)—and/or multi (indels)—nucleotide variants, copy number variants (CNVs), and other structural variants (SVs), a few filtering, and quality checks on the SAM/BAM files are performed. These include removal of duplicates and reads mapped at multiple locations in the genome, realigning reads with known Indels, and recalibrating base quality scores with respect to the known SNVs. Once the SAM/BAM files are checked for quality, the files are used for variant calling. Although there are

Table 1 Different types of cancer data generated from a typical Illumina sequencing instrument and their descriptions.

Data types	Description
Raw data	• Multiple instruments, from multiple vendors and produced by multiple scientists • Produced as image data and never leaves the primary instrument that drives the sequencing instruments • Image data (in multi-terabytes) are automatically discarded after the initial QC • Image files are converted to base call files (BCL) after the initial QC and transferred to a server for further analyses
Primary data	• Usable data that is derived from raw data and serves as the first-level primary data for all subsequent analyses • Usually in the form of fastq files • Usually kept for 3–5 years depending on the project duration and complexity before being archived • A single fastq file size can vary anywhere between 5–15 Gb (for a 50 MB human exome at 50–100X) to 200–350 Gb (for a 30–50X whole human genome)
Secondary data	• The fastq files are aligned to a reference genome and kept as aligned file (SAM/BAM format) • A single SAM/BAM file can vary in size (2–5 GB in size for a single 50 MB human exome at 50–100X).
Tertiary data	• Files produced after the variant calls are made (in most cases VCF files) • Usually vary in size (40–250 Mb for a 50 MB human exome data with 50–100X coverage) • Kept for selected project-specific files for up to a year and then archived
Final results	• Amalgamated with other data types from multiple instruments • Used in publication • Between 10Mb - Gb in size depending on the data type for each sample • Kept for a very long time

multiple tools for calling variants, the widely used popular one is the genome analysis toolkit (GATK) [21, 22] developed at the Broad Institute, USA. GATK implements variant quality score recalibration and posterior probability calculations to minimize the false positive rates in the pool of variants called [22]. Variants are stored in a file format called variant call format (VCF), which is used by various secondary and tertiary analysis tools. Another commonly used file format for cancer data analysis is called mutation annotation format (MAF), initially made to analyze data coming out from the cancer genome atlas (TCGA) consortium. The MAF format lists all the mutations and stores much more information about the variants and alignment than the VCF files.

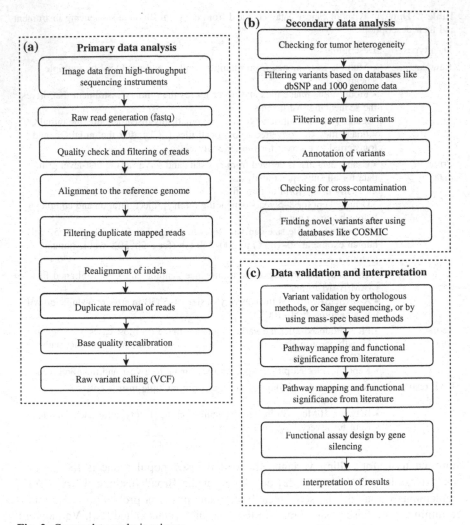

Fig. 2 Cancer data analysis schema

6 Secondary Data Analysis

Before secondary analysis, usually the PASS variants produced by GATK (standard call confidence >= 50) within specific genomic bait (used in exome or gene panels) are filtered and taken for further use. Tumor-specific variants are detected by filtering out the variants found in its corresponding/paired normal sample. During this process of calling tumor-specific variants, sequencing reads representing a particular variant in a tumor sample that have no corresponding reads in matched normal in the same location are ignored (using the callable filter of the variant caller). Then common SNPs (found in the normal population) are filtered out using a list of variants found in the databases like dbSNP and 1000 genome project. Therefore, only variants that are represented by sequencing reads both in tumor and its matched normal samples are considered. Optimization methods/workflows have been designed to analytically access the best combination of tools (both alignment and variant calling) to increase the sensitivity of variant detection [23]. The sensitivity of the alignment and variant calling tools are usually assessed by a set of metrics like aligner and variant caller-specific base quality plots of the variants called, transition/transversion (Ti/Tv) ratios, and SNP rediscovery rate using microarrays [23]. Further cross-contamination in tumor samples are assessed using tools like ContEst [24]. Searching variants against the known cancer-specific variants in databases like COSMIC [25–27] is the first step to find out whether the variant/gene is unique/novel or have been found in the same or other cancer types previously. There are cancer-specific tools to perform annotation and functional analysis. The common annotation tools are ANNOVAR [28] and VEP [29]. CRAVAT [30] provides predictive scores for different types of variants (both somatic and germline) and annotations from published literature and databases. It uses a specific cancer database with the CHASM analysis option. Genes with a CHASM score of a certain value are considered significant for comparison with other functional analyses. IntoGen [31], MutSigCV [32], and MuSiC2 [33] are other tools that are used for annotation and functional analyses of somatic variants.

7 Data Validation, Visualization, and Interpretation

Once annotated, the cancer-specific genes are validated in the same discovery set and also using a larger set of validation samples. Validation is largely done using either an orthologous sequencing method/chemistry, mass-spec-based mutation detection methods, and/or using Sanger sequencing technique. Finally the validated variants are mapped to pathways using tools like Graphite Web [34] that employs both the topological and multivariate pathway analyses with an interactive network for data visualizations. Once the network of genes is obtained, the interactions are drawn using tools like CytoScape [35–37]. Variants can also be visualized by using Circos [38], a cancer-specific portal like cbio portal [39] or with a viewer like the integrative genomics viewer (IGV) [40]. Finally, the genes that are altered in a specific cancer tissue are validated using functional screening methods using specific gene knockouts to understand their function and relationship with other genes.

8 High-Throughput Data on Human Cancers

The current projects on cancer genomics are aimed to produce a large amount of sequence information as primary output and information on variant data (somatic mutations, insertions and deletions, copy number variations and other structural variations in the genome). In order to analyze the large amount of data, high-performance compute clusters (HPC) with large memory and storage capacity are required. Additionally, higher frequency, high-throughput multi-core chips along with the ability to do high-volume data analysis in memory are often required. Due to the sheer number of files, and not just the size of the files, that need to be processed, the read/write capability is an important parameter for sequence analysis. For effective storage and analysis of sequencing and related metadata, network access storage systems, providing file-level access, are recommended. Additionally, there is a need for an effective database for data organization for easy access, management, and data update. Several data portals, primarily made by the large consortia are developed. Prominent among them are: The Cancer Genome Atlas (TCGA, https://tcga-data.nci.nih.gov/tcga/) data portal; cbio data portal [39] (developed at the Memorial Sloan-Kettering Cancer Center, http://www.cbioportal. org); the International Cancer Genome Consortium (ICGC) data portal (https://dcc. icgc.org); and the Sanger Institute's Catalogue of Somatic Mutations in Cancer (COSMIC) database [25] portal (http://cancer.sanger.ac.uk/cosmic).

Although biological databases are created using many different platforms, the most common among them are MySQL and Oracle. MySQL is more popular database because of its open source. Although the consortia-led efforts (like TCGA and ICGC) have resulted in large and comprehensive databases covering most cancer types, the sites are not user-friendly and do not accept external data for integration and visualization. Therefore, efforts like cbio portal (http://www. cbioportal.org) are required to integrate data and user-friendly data search and retrieval. However, such efforts have to balance keeping in mind the cost and time required versus usability and additional value addition from the new database. The common databases use software systems known as Relational Database Management Systems (RDBMS) that use SQL (Structured Query Language) for querying and maintaining the databases. MySQL is a widely used open source RDBMS. Although most biological database uses MySQL or other RDBMS, it has its limitations as far as large data is concerned. First, big data is assumed to come in structured, semi-structured, and unstructured manner. Second, traditional SQL databases and other RDBMS lack ability to scale out a requirement for databases containing large amount of data. Third, RDBMS cannot scale out with inexpensive hardware. All these make RDBMS unsuitable for large data uses. This is primarily filled by other databases like NoSQL that are document-oriented graph databases that are non-relational, friendly to HPC environment, schema-less, and built to scale [41]. One of the important parameters in a database is the ability to take care of future increase in data size and complexity (Fig. 3), therefore having an ability to scale in both these parameters. Although it is a good idea to think of databases that

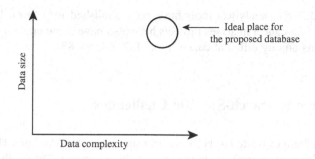

Fig. 3 Two important parameters of big data and the place for an ideal database

have the ability to scale out, and accommodate variety and volume of future data increase, due to simplicity and ease of use, most small labs stick with MySQL database that uses variety of data, commonly used middleware and web server, and browser for data retrieval and visualization.

9 Large-Scale Cancer Genome Projects

Advances in technology have fuelled interest in the cancer research community that has resulted in several large publicly funded consortia-based efforts to catalogue changes in primary tumors of various types. Some of the notable and prominent efforts in this direction are, The Cancer Genome Atlas (TCGA) project (http://www.cancergenome.nih.gov/), the International Cancer Genome Consortium (ICGC) project (https://icgc.org) [42], the Cancer Genome Project (http://www.sanger.ac.uk/genetics/CGP/), and the Therapeutically applicable Research to Generate Effective Treatments (http://target.cancer.gov/) project. The National Cancer Institute (NCI) and the National Human Genome Research Institute (NHGRI) of USA initially launched the TCGA as a pilot project in 2006 even before the first human resequencing work using HTS platforms was published. The TCGA effort plans to produce a comprehensive understanding of the molecular basis of cancer and currently has grown to include samples from more than 11,000 patients across 33 different cancer types. The ICGC is an international consortium that plans to obtain a comprehensive description of various molecular changes (genomic, transcriptomic and epigenomic) in 50 different tumor types and/or subtypes. ICGC currently has participants from 18 countries studying cancer samples from more than 12000 donors on 21 tumor types. All the consortia projects are producing substantial resource for the wider cancer research community.

Till date, HTS data on several cancer types have been generated and data analysis confirmed the presence of somatic mutations in important genes, significant changes in gene/miRNA expression, hyper- and hypo-methylation in gene promoters, and structural variations in the cancer genomes [32, 43–66]. Additionally, comparative analysis of different analytical tools have been published for cancer data analysis [23, 67–83]. Pan-cancer analyses projects have also have come up with specific and shared regions among different cancer types [32, 65, 84–88].

10 Cancer Research-Specific Challenges

There are challenges related to HTS assays using tumor tissues. For a HTS assay to have clinical utility, several challenges need to be overcome. The challenges can be clinical, technical, biological, statistical, regulatory, and market-related and are outlined in Table 2.

Clinical challenges: First among the clinical challenges is related to sample quantity. For retrospective studies to be meaningful, assays must be robust to use nucleic acids derived from formalin-fixed paraffin-embedded (FFPE) tissues. Tissue sections extracted are not often big to yield sufficient quantity of nucleic acids that can be used for sequencing, and validation studies even with the newer assays that

Table 2 Challenges of making high-throughput assays, especially sequencing-based assays, meaningful in clinics

Type of challenge	Issues
Clinical	• Sample quantity from tissue biopsies • Sample quality (FFPE tissue, chemical-crosslinking, etc.) • Tumor heterogeneity
Biological	• Background somatic mutation (needle in a haystack problem) • Varying mutation rate • Finding the right matched normal • Lack of proper controls for data assessment
Technical	• Sequencing error rate • Repeats in the genome • Read length • Stretch of homopolymers in the genome • Regions of low-complexity in the genome • Nucleotide composition
Statistical	• Large sample number requirement to catch somatic mutations at low frequency
Regulatory	• Lack of standards for clinical evaluation • Lack of controls to judge sequencing and variant calling error and accuracy • Lack of proper regulatory guidelines
Market-related	• Price • Competition • Acceptability

use only tens of nanograms of nucleic acids as starting material. Even if one manages to get enough nucleic acids from the FFPE tissues, the quality of nucleic acids extracted is not the best of quality and is often fragmented. Additionally, chemical modifications like the presence of cross-links and depurination and the presence of certain impurities in the FFPE-extracted DNA make them less amenable to alterations required for high-throughput assays. Therefore, FFPE-extracted DNA can have a stronger influence on the HTS assays. Cancer tissues are heterogeneous [32, 89] and in certain cases extremely heterogeneous (for example in pancreatic adenocarcinoma) that cautions overinterpreting HTS data from a lump of tumor tissue as shown in metastatic renal-cell carcinoma [90]. Therefore, in heterogenous tumors, the mutational burden may be underestimated. Studying such intra-tumor heterogeneity may aid the case for combination therapeutic approaches in cancer [91]. Analytical methods have been devised in the past to detect tumor heterogeneity [73, 92, 93].

Biological challenges: The next challenge is biological where finding somatic mutations, especially, those present at very low frequency, among the sea of normal background is really difficult. The use of a matched normal sample for cancer sequencing is essential to find somatic variants but, at times, the matched normal tissue might be hard to get and therefore, variants found in lymphocytes from the same patients are often used as normal samples. Another problem in sequencing tumor tissue DNA is cross-contamination. Analytical tools have been developed to detect the level of cross-contamination in tumor tissues from both sequencing and array data [24, 94]. To overcome both the heterogeneity and the cross-contamination issue, the best way is to perform DNA/RNA sequencing derived from a single tumor cell. Single-cell genomics is likely to help and improve detection, progression, and prediction of therapeutic efficacy of cancer [95]. Several reports have been published on single-cell sequencing of different cancers and analytical tool development to analyze data from a single tumor cell [96–108]. Although the problem of heterogeneity is overcome with single-cell sequencing, the fundamental questions may still linger, i.e., how many single cells have to be sequenced and if the signature is different in different single tumor cells. Additionally, there are limitations to the current protocols for isolation of single tumor cells and the inaccuracies involved in whole genome amplification of genomic DNA derived from a single cell. Therefore, capturing minute amounts of genetic material and amplifying them remain as one the greatest challenges in single cell genomics [109, 110].

Technical challenges: The third type of challenge is related to technical issues with current generation of sequencing instruments. Depending on the instrument in use, there could be an issue related to high error rate, length of the read, homopolymer stretches, and GC-rich regions in the genome. Additionally, accurate sequencing and assembling correct haplotype structures for certain regions of the genome, like the human leukocyte antigen (HLA) region, are challenging due to shorter read lengths generated in second generation DNA sequencers, presence of polymorphic exons and pseudogenes, and repeat rich region.

Statistical challenges: One of the biggest challenges to find driver mutations in cancer is related to sample number. Discovering rare driver mutations in cancer is extremely challenging, especially when sample numbers are not adequate. This, so-called, "the long tail phenomenon" is quite common in many of the cancer genome sequencing studies. Discovering rare driver mutations (found at 2 % frequency or lower) requires sequencing a large number of samples. For example, in head and neck cancer, imputations have shown that it will take 2000 tumor:normal samples to be sequenced at 90 % power in 90 % of the genes to find somatic variants present at 2 % frequency or higher [43].

Regulatory and other challenges: In order for cancer personalized medicine to become a reality, proper regulatory and policy framework need to be in place. Issues around how to deal with germline changes along with strict and proper assay and technical controls/standards are needed to be in place to assess biological, clinical, and technical accuracy and authenticity. A great beginning in this direction has already been made by the genome in a bottle consortium (https://sites.stanford.edu/abms/giab) hosted by the National Institute of Standards and Technology of the USA that has come up with reference materials (reference standards, reference methods, and reference data) to be used in sequencing. Finally, in order for cutting edge genomic tests to become a reality, collaboration and cooperation between academic centers and industry are absolutely necessary [111]. Additionally, acceptability criteria and proper pricing control mechanism(s) need to be in place by the government. This is necessary for countries like India where genomic tests are largely unregulated.

11   Conclusion

Cancer research has changed since the introduction of technologies like DNA microarray and high-throughput sequencing. It is now possible to get a genome-wide view on a particular tumor rather than looking at a handful of genes. The biggest challenge for finding actionable variants in cancer remains at the level of data analysis and understanding of their functional importance. Recent demonstrations [112–115] of gene editing systems like CRISPR-Cas9 in understanding the function of cancer-related genes and their role(s) in carcinogenesis and metastasis will play a big role in the future. Further, high-throughput sequencing technology can be used to providing information on individual cancer regulome by integrating information on genetic variants, transcript variants, regulatory proteins binding to DNA and RNA, DNA and protein methylation, and metabolites. Finally, for big data to bear fruits in cancer diagnosis, prognosis, and treatment, processes like; simplified data analytics platforms; accurate sequencing chemistry; standards for measuring clinical accuracy, precision and sensitivity; proper country-specific regulatory guidelines and stringent yet ethical framework against data misuse; need to be in place [111].

Acknowledgments Research in Ganit Labs, Bio-IT Centre is funded by grants from the Government of India agencies (Department of Electronics and Information Technology; Department of Biotechnology; Department of Science and Technology; and the Council of Scientific and Industrial Research) and Department of Information Technology, Biotechnology and Science & Technology, Government of Karnataka, India. I thank Saurabh Gupta for helping in making Fig. 2, and Saurabh Gupta and Neeraja Krishnan for critically reading the manuscript. Ganit Labs is an initiative of Institute of Bioinformatics and Applied Biotechnology and Strand Life Sciences, both located in Bangalore, India.

References

1. Maxam AM, Gilbert W (1977) A new method for sequencing DNA. Proc Natl Acad Sci USA 74:560–564
2. Sanger F, Nicklen S, Coulson AR (1977) DNA sequencing with chain-terminating inhibitors. Proc Natl Acad Sci USA 74:5463–5467
3. Venter JC, Adams MD, Myers EW, Li PW, Mural RJ, Sutton GG, Smith HO, Yandell M, Evans CA, Holt RA et al (2001) The sequence of the human genome. Science 291:1304–1351
4. Lander ES, Linton LM, Birren B, Nusbaum C, Zody MC, Baldwin J, Devon K, Dewar K, Doyle M, FitzHugh W et al (2001) Initial sequencing and analysis of the human genome. Nature 409:860–921
5. Metzker ML (2010) Sequencing technologies - the next generation. Nat Rev Genet 11:31–46
6. Wheeler DA, Srinivasan M, Egholm M, Shen Y, Chen L, McGuire A, He W, Chen YJ, Makhijani V, Roth GT et al (2008) The complete genome of an individual by massively parallel DNA sequencing. Nature 452:872–876
7. Bentley DR, Balasubramanian S, Swerdlow HP, Smith GP, Milton J, Brown CG, Hall KP, Evers DJ, Barnes CL, Bignell HR et al (2008) Accurate whole human genome sequencing using reversible terminator chemistry. Nature 456:53–59
8. Homer N, Merriman B, Nelson SF (2009) BFAST: an alignment tool for large scale genome resequencing. PLoS ONE 4:e7767
9. Ning Z, Cox AJ, Mullikin JC (2001) SSAHA: a fast search method for large DNA databases. Genome Res 11:1725–1729
10. SMALT [http://www.sanger.ac.uk/resources/software/smalt/]
11. Lunter G, Goodson M (2011) Stampy: a statistical algorithm for sensitive and fast mapping of Illumina sequence reads. Genome Res 21:936–939
12. Novoalign (www.novocraft.com)
13. Langmead B (2010) Aligning short sequencing reads with Bowtie. Curr Protoc Bioinform., Chap 11:Unit 11–17
14. Langmead B, Salzberg SL (2012) Fast gapped-read alignment with Bowtie 2. Nat Methods 9:357–359
15. Li H, Durbin R (2009) Fast and accurate short read alignment with Burrows-Wheeler transform. Bioinformatics 25:1754–1760
16. Liu Y, Schmidt B, Maskell DL (2012) CUSHAW: a CUDA compatible short read aligner to large genomes based on the Burrows-Wheeler transform. Bioinformatics 28:1830–1837
17. Klus P, Lam S, Lyberg D, Cheung MS, Pullan G, McFarlane I, Yeo G, Lam BY (2012) BarraCUDA—a fast short read sequence aligner using graphics processing units. BMC Res Notes 5:27
18. Gupta S, Choudhury S, Panda B (2014) MUSIC: A hybrid-computing environment for Burrows-Wheeler alignment for massive amount of short read sequence data. MECBME 2014 (indexed in IEEE Xplore)

19. Schatz MC, Trapnell C, Delcher AL, Varshney A (2007) High-throughput sequence alignment using graphics processing units. BMC Bioinform 8:474

20. Li H, Handsaker B, Wysoker A, Fennell T, Ruan J, Homer N, Marth G, Abecasis G, Durbin R (2009) The sequence alignment/map format and SAMtools. Bioinformatics 25:2078–2079

21. McKenna A, Hanna M, Banks E, Sivachenko A, Cibulskis K, Kernytsky A, Garimella K, Altshuler D, Gabriel S, Daly M, DePristo MA (2010) The Genome Analysis Toolkit: a MapReduce framework for analyzing next-generation DNA sequencing data. Genome Res 20:1297–1303

22. DePristo MA, Banks E, Poplin R, Garimella KV, Maguire JR, Hartl C, Philippakis AA, del Angel G, Rivas MA, Hanna M et al (2011) A framework for variation discovery and genotyping using next-generation DNA sequencing data. Nat Genet 43:491–498

23. Pattnaik S, Vaidyanathan S, Pooja DG, Deepak S, Panda B (2012) Customisation of the exome data analysis pipeline using a combinatorial approach. PLoS ONE 7:e30080

24. Cibulskis K, McKenna A, Fennell T, Banks E, DePristo M, Getz G (2011) ContEst: estimating cross-contamination of human samples in next-generation sequencing data. Bioinformatics 27:2601–2602

25. Forbes SA, Beare D, Gunasekaran P, Leung K, Bindal N, Boutselakis H, Ding M, Bamford S, Cole C, Ward S et al (2015) COSMIC: exploring the world's knowledge of somatic mutations in human cancer. Nucleic Acids Res 43:D805–D811

26. Forbes SA, Bindal N, Bamford S, Cole C, Kok CY, Beare D, Jia M, Shepherd R, Leung K, Menzies A et al (2011) COSMIC: mining complete cancer genomes in the Catalogue of Somatic Mutations in Cancer. Nucleic Acids Res 39:D945–D950

27. Forbes SA, Tang G, Bindal N, Bamford S, Dawson E, Cole C, Kok CY, Jia M, Ewing R, Menzies A et al (2010) COSMIC (the Catalogue of Somatic Mutations in Cancer): a resource to investigate acquired mutations in human cancer. Nucleic Acids Res 38:D652–D657

28. Wang K, Li M, Hakonarson H (2010) ANNOVAR: functional annotation of genetic variants from high-throughput sequencing data. Nucleic Acids Res 38:e164

29. Yourshaw M, Taylor SP, Rao AR, Martin MG, Nelson SF (2015) Rich annotation of DNA sequencing variants by leveraging the Ensembl Variant Effect Predictor with plugins. Brief Bioinform 16:255–264

30. Douville C, Carter H, Kim R, Niknafs N, Diekhans M, Stenson PD, Cooper DN, Ryan M, Karchin R (2013) CRAVAT: cancer-related analysis of variants toolkit. Bioinformatics 29:647–648

31. Gundem G, Perez-Llamas C, Jene-Sanz A, Kedzierska A, Islam A, Deu-Pons J, Furney SJ, Lopez-Bigas N (2010) IntOGen: integration and data mining of multidimensional oncogenomic data. Nat Methods 7:92–93

32. Lawrence MS, Stojanov P, Polak P, Kryukov GV, Cibulskis K, Sivachenko A, Carter SL, Stewart C, Mermel CH, Roberts SA et al (2013) Mutational heterogeneity in cancer and the search for new cancer-associated genes. Nature 499:214–218

33. Dees ND: MuSiC2. 2015

34. Sales G, Calura E, Martini P, Romualdi C (2013) Graphite Web: Web tool for gene set analysis exploiting pathway topology. Nucleic Acids Res 41:W89–W97

35. Lopes CT, Franz M, Kazi F, Donaldson SL, Morris Q, Bader GD (2010) Cytoscape Web: an interactive web-based network browser. Bioinformatics 26:2347–2348

36. Cline MS, Smoot M, Cerami E, Kuchinsky A, Landys N, Workman C, Christmas R, Avila-Campilo I, Creech M, Gross B et al (2007) Integration of biological networks and gene expression data using Cytoscape. Nat Protoc 2:2366–2382

37. Shannon P, Markiel A, Ozier O, Baliga NS, Wang JT, Ramage D, Amin N, Schwikowski B, Ideker T (2003) Cytoscape: a software environment for integrated models of biomolecular interaction networks. Genome Res 13:2498–2504

38. Krzywinski M, Schein J, Birol I, Connors J, Gascoyne R, Horsman D, Jones SJ, Marra MA (2009) Circos: an information aesthetic for comparative genomics. Genome Res 19:1639–1645

39. Cerami E, Gao J, Dogrusoz U, Gross BE, Sumer SO, Aksoy BA, Jacobsen A, Byrne CJ, Heuer ML, Larsson E et al (2012) The cBio cancer genomics portal: an open platform for exploring multidimensional cancer genomics data. Cancer Discov 2:401–404

40. Robinson JT, Thorvaldsdottir H, Winckler W, Guttman M, Lander ES, Getz G, Mesirov JP (2011) Integrative genomics viewer. Nat Biotechnol 29:24–26

41. Hu H, Wen Y, Chua TS, Li X (2014) Toward scalable systems for big data analytics: a technology tutorial. IEEE Access 2:652–687

42. Hudson TJ, Anderson W, Artez A, Barker AD, Bell C, Bernabe RR, Bhan MK, Calvo F, Eerola I, Gerhard DS et al (2010) International network of cancer genome projects. Nature 464:993–998

43. Lawrence MS, Stojanov P, Mermel CH, Robinson JT, Garraway LA, Golub TR, Meyerson M, Gabriel SB, Lander ES, Getz G (2014) Discovery and saturation analysis of cancer genes across 21 tumour types. Nature 505:495–501

44. Stephens PJ, McBride DJ, Lin ML, Varela I, Pleasance ED, Simpson JT, Stebbings LA, Leroy C, Edkins S, Mudie LJ et al (2009) Complex landscapes of somatic rearrangement in human breast cancer genomes. Nature 462:1005–1010

45. van Haaften G, Dalgliesh GL, Davies H, Chen L, Bignell G, Greenman C, Edkins S, Hardy C, O'Meara S, Teague J et al (2009) Somatic mutations of the histone H3K27 demethylase gene UTX in human cancer. Nat Genet 41:521–523

46. Pleasance ED, Cheetham RK, Stephens PJ, McBride DJ, Humphray SJ, Greenman CD, Varela I, Lin ML, Ordonez GR, Bignell GR et al (2010) A comprehensive catalogue of somatic mutations from a human cancer genome. Nature 463:191–196

47. Pleasance ED, Stephens PJ, O'Meara S, McBride DJ, Meynert A, Jones D, Lin ML, Beare D, Lau KW, Greenman C et al (2010) A small-cell lung cancer genome with complex signatures of tobacco exposure. Nature 463:184–190

48. Papaemmanuil E, Cazzola M, Boultwood J, Malcovati L, Vyas P, Bowen D, Pellagatti A, Wainscoat JS, Hellstrom-Lindberg E, Gambacorti-Passerini C et al (2011) Somatic SF3B1 mutation in myelodysplasia with ring sideroblasts. N Engl J Med 365:1384–1395

49. Puente XS, Pinyol M, Quesada V, Conde L, Ordonez GR, Villamor N, Escaramis G, Jares P, Bea S, Gonzalez-Diaz M et al (2011) Whole-genome sequencing identifies recurrent mutations in chronic lymphocytic leukaemia. Nature 475:101–105

50. Stephens PJ, Greenman CD, Fu B, Yang F, Bignell GR, Mudie LJ, Pleasance ED, Lau KW, Beare D, Stebbings LA et al (2011) Massive genomic rearrangement acquired in a single catastrophic event during cancer development. Cell 144:27–40

51. Varela I, Tarpey P, Raine K, Huang D, Ong CK, Stephens P, Davies H, Jones D, Lin ML, Teague J et al (2011) Exome sequencing identifies frequent mutation of the SWI/SNF complex gene PBRM1 in renal carcinoma. Nature 469:539–542

52. Greenman CD, Pleasance ED, Newman S, Yang F, Fu B, Nik-Zainal S, Jones D, Lau KW, Carter N, Edwards PA et al (2012) Estimation of rearrangement phylogeny for cancer genomes. Genome Res 22:346–361

53. Nik-Zainal S, Alexandrov LB, Wedge DC, Van Loo P, Greenman CD, Raine K, Jones D, Hinton J, Marshall J, Stebbings LA et al (2012) Mutational processes molding the genomes of 21 breast cancers. Cell 149:979–993

54. Stephens PJ, Tarpey PS, Davies H, Van Loo P, Greenman C, Wedge DC, Nik-Zainal S, Martin S, Varela I, Bignell GR et al (2012) The landscape of cancer genes and mutational processes in breast cancer. Nature 486:400–404

55. Wang L, Tsutsumi S, Kawaguchi T, Nagasaki K, Tatsuno K, Yamamoto S, Sang F, Sonoda K, Sugawara M, Saiura A et al (2012) Whole-exome sequencing of human pancreatic cancers and characterization of genomic instability caused by MLH1 haploinsufficiency and complete deficiency. Genome Res 22:208–219

56. Cancer Genome Atlas N (2015) Comprehensive genomic characterization of head and neck squamous cell carcinomas. Nature 517:576–582

57. India Project Team of the International Cancer Genome C (2013) Mutational landscape of gingivo-buccal oral squamous cell carcinoma reveals new recurrently-mutated genes and molecular subgroups. Nat Commun 4:2873

58. Barbieri CE, Baca SC, Lawrence MS, Demichelis F, Blattner M, Theurillat JP, White TA, Stojanov P, Van Allen E, Stransky N et al (2012) Exome sequencing identifies recurrent SPOP, FOXA1 and MED12 mutations in prostate cancer. Nat Genet 44:685–689

59. Van Allen EM, Wagle N, Stojanov P, Perrin DL, Cibulskis K, Marlow S, Jane-Valbuena J, Friedrich DC, Kryukov G, Carter SL et al (2014) Whole-exome sequencing and clinical interpretation of formalin-fixed, paraffin-embedded tumor samples to guide precision cancer medicine. Nat Med 20:682–688

60. Wang L, Lawrence MS, Wan Y, Stojanov P, Sougnez C, Stevenson K, Werner L, Sivachenko A, DeLuca DS, Zhang L et al (2011) SF3B1 and other novel cancer genes in chronic lymphocytic leukemia. N Engl J Med 365:2497–2506

61. Craig DW, O'Shaughnessy JA, Kiefer JA, Aldrich J, Sinari S, Moses TM, Wong S, Dinh J, Christoforides A, Blum JL et al (2013) Genome and transcriptome sequencing in prospective metastatic triple-negative breast cancer uncovers therapeutic vulnerabilities. Mol Cancer Ther 12:104–116

62. Beltran H, Rickman DS, Park K, Chae SS, Sboner A, MacDonald TY, Wang Y, Sheikh KL, Terry S, Tagawa ST et al (2011) Molecular characterization of neuroendocrine prostate cancer and identification of new drug targets. Cancer Discov 1:487–495

63. Drier Y, Lawrence MS, Carter SL, Stewart C, Gabriel SB, Lander ES, Meyerson M, Beroukhim R, Getz G (2013) Somatic rearrangements across cancer reveal classes of samples with distinct patterns of DNA breakage and rearrangement-induced hypermutability. Genome Res 23:228–235

64. Eswaran J, Horvath A, Godbole S, Reddy SD, Mudvari P, Ohshiro K, Cyanam D, Nair S, Fuqua SA, Polyak K et al (2013) RNA sequencing of cancer reveals novel splicing alterations. Sci Rep 3:1689

65. Kandoth C, McLellan MD, Vandin F, Ye K, Niu B, Lu C, Xie M, Zhang Q, McMichael JF, Wyczalkowski MA et al (2013) Mutational landscape and significance across 12 major cancer types. Nature 502:333–339

66. Wu X, Cao W, Wang X, Zhang J, Lv Z, Qin X, Wu Y, Chen W (2013) TGM3, a candidate tumor suppressor gene, contributes to human head and neck cancer. Mol Cancer 12:151

67. Merid SK, Goranskaya D, Alexeyenko A (2014) Distinguishing between driver and passenger mutations in individual cancer genomes by network enrichment analysis. BMC Bioinform 15:308

68. Layer RM, Chiang C, Quinlan AR, Hall IM (2014) LUMPY: a probabilistic framework for structural variant discovery. Genome Biol 15:R84

69. Dietlein F, Eschner W (2014) Inferring primary tumor sites from mutation spectra: a meta-analysis of histology-specific aberrations in cancer-derived cell lines. Hum Mol Genet 23:1527–1537

70. Cole C, Krampis K, Karagiannis K, Almeida JS, Faison WJ, Motwani M, Wan Q, Golikov A, Pan Y, Simonyan V, Mazumder R (2014) Non-synonymous variations in cancer and their effects on the human proteome: workflow for NGS data biocuration and proteome-wide analysis of TCGA data. BMC Bioinform 15:28

71. Wittler R (2013) Unraveling overlapping deletions by agglomerative clustering. BMC Genom 14(Suppl 1):S12

72. Trifonov V, Pasqualucci L, Dalla Favera R, Rabadan R (2013) MutComFocal: an integrative approach to identifying recurrent and focal genomic alterations in tumor samples. BMC Syst Biol 7:25

73. Oesper L, Mahmoody A, Raphael BJ (2013) THetA: inferring intra-tumor heterogeneity from high-throughput DNA sequencing data. Genome Biol 14:R80

74. Hansen NF, Gartner JJ, Mei L, Samuels Y, Mullikin JC (2013) Shimmer: detection of genetic alterations in tumors using next-generation sequence data. Bioinformatics 29:1498–1503

75. Hamilton MP, Rajapakshe K, Hartig SM, Reva B, McLellan MD, Kandoth C, Ding L, Zack TI, Gunaratne PH, Wheeler DA et al (2013) Identification of a pan-cancer oncogenic microRNA superfamily anchored by a central core seed motif. Nat Commun 4:2730

76. Chen Y, Yao H, Thompson EJ, Tannir NM, Weinstein JN, Su X (2013) VirusSeq: software to identify viruses and their integration sites using next-generation sequencing of human cancer tissue. Bioinformatics 29:266–267

77. Mosen-Ansorena D, Telleria N, Veganzones S, De la Orden V, Maestro ML, Aransay AM (2014) seqCNA: an R package for DNA copy number analysis in cancer using high-throughput sequencing. BMC Genom 15:178

78. Li Y, Xie X (2014) Deconvolving tumor purity and ploidy by integrating copy number alterations and loss of heterozygosity. Bioinformatics 30:2121–2129

79. Kendall J, Krasnitz A (2014) Computational methods for DNA copy-number analysis of tumors. Methods Mol Biol 1176:243–259

80. Krishnan NM, Gaur P, Chaudhary R, Rao AA, Panda B (2012) COPS: a sensitive and accurate tool for detecting somatic Copy Number Alterations using short-read sequence data from paired samples. PLoS ONE 7:e47812

81. Van Allen EM, Wagle N, Levy MA (2013) Clinical analysis and interpretation of cancer genome data. J Clin Oncol 31:1825–1833

82. Lahti L, Schafer M, Klein HU, Bicciato S, Dugas M (2013) Cancer gene prioritization by integrative analysis of mRNA expression and DNA copy number data: a comparative review. Brief Bioinform 14:27–35

83. Lee LA, Arvai KJ, Jones D (2015) Annotation of sequence variants in cancer samples: processes and pitfalls for routine assays in the clinical laboratory. J Mol Diagn

84. Weinstein JN, Collisson EA, Mills GB, Shaw KR, Ozenberger BA, Ellrott K, Shmulevich I, Sander C, Stuart JM (2013) The cancer genome Atlas Pan-cancer analysis project. Nat Genet 45:1113–1120

85. Zack TI, Schumacher SE, Carter SL, Cherniack AD, Saksena G, Tabak B, Lawrence MS, Zhang CZ, Wala J, Mermel CH et al (2013) Pan-cancer patterns of somatic copy number alteration. Nat Genet 45:1134–1140

86. Gross AM, Orosco RK, Shen JP, Egloff AM, Carter H, Hofree M, Choueiri M, Coffey CS, Lippman SM, Hayes DN et al (2014) Multi-tiered genomic analysis of head and neck cancer ties TP53 mutation to 3p loss. Nat Genet 46:939–943

87. Pan-cancer initiative finds patterns of drivers (2013) Cancer Discov 3:1320

88. Taking pan-cancer analysis global (2013) Nat Genet 45:1263

89. Russnes HG, Navin N, Hicks J, Borresen-Dale AL (2011) Insight into the heterogeneity of breast cancer through next-generation sequencing. J Clin Invest 121:3810–3818

90. Gerlinger M, Rowan AJ, Horswell S, Larkin J, Endesfelder D, Gronroos E, Martinez P, Matthews N, Stewart A, Tarpey P et al (2012) Intratumor heterogeneity and branched evolution revealed by multiregion sequencing. N Engl J Med 366:883–892

91. Swanton C (2012) Intratumor heterogeneity: evolution through space and time. Cancer Res 72:4875–4882

92. Oesper L, Satas G, Raphael BJ (2014) Quantifying tumor heterogeneity in whole-genome and whole-exome sequencing data. Bioinformatics 30:3532–3540

93. Hajirasouliha I, Mahmoody A, Raphael BJ (2014) A combinatorial approach for analyzing intra-tumor heterogeneity from high-throughput sequencing data. Bioinformatics 30:i78–i86

94. Jun G, Flickinger M, Hetrick KN, Romm JM, Doheny KF, Abecasis GR, Boehnke M, Kang HM (2012) Detecting and estimating contamination of human DNA samples in sequencing and array-based genotype data. Am J Hum Genet 91:839–848

95. Navin N, Hicks J (2011) Future medical applications of single-cell sequencing in cancer. Genome Med 3:31

96. Ji C, Miao Z, He X (2015) A simple strategy for reducing false negatives in calling variants from single-cell sequencing data. PLoS ONE 10:e0123789

97. Yu C, Yu J, Yao X, Wu WK, Lu Y, Tang S, Li X, Bao L, Li X, Hou Y et al (2014) Discovery of biclonal origin and a novel oncogene SLC12A5 in colon cancer by single-cell sequencing. Cell Res 24:701–712

98. Ting DT, Wittner BS, Ligorio M, Vincent Jordan N, Shah AM, Miyamoto DT, Aceto N, Bersani F, Brannigan BW, Xega K et al (2014) Single-cell RNA sequencing identifies extracellular matrix gene expression by pancreatic circulating tumor cells. Cell Rep 8:1905–1918

99. Kim KI, Simon R (2014) Using single cell sequencing data to model the evolutionary history of a tumor. BMC Bioinform 15:27

100. Xu Y, Hu H, Zheng J, Li B (2013) Feasibility of whole RNA sequencing from single-cell mRNA amplification. Genet Res Int 2013:724124

101. Voet T, Kumar P, Van Loo P, Cooke SL, Marshall J, Lin ML, Zamani Esteki M, Van der Aa N, Mateiu L, McBride DJ et al (2013) Single-cell paired-end genome sequencing reveals structural variation per cell cycle. Nucleic Acids Res 41:6119–6138

102. Korfhage C, Fisch E, Fricke E, Baedker S, Loeffert D (2013) Whole-genome amplification of single-cell genomes for next-generation sequencing. Curr Protoc Mol Biol 104:Unit 7–14

103. Geurts-Giele WR, Dirkx-van der Velden AW, Bartalits NM, Verhoog LC, Hanselaar WE, Dinjens WN (2013) Molecular diagnostics of a single multifocal non-small cell lung cancer case using targeted next generation sequencing. Virchows Arch 462:249–254

104. Xu X, Hou Y, Yin X, Bao L, Tang A, Song L, Li F, Tsang S, Wu K, Wu H et al (2012) Single-cell exome sequencing reveals single-nucleotide mutation characteristics of a kidney tumor. Cell 148:886–895

105. Li Y, Xu X, Song L, Hou Y, Li Z, Tsang S, Li F, Im KM, Wu K, Wu H et al (2012) Single-cell sequencing analysis characterizes common and cell-lineage-specific mutations in a muscle-invasive bladder cancer. Gigascience 1:12

106. Hou Y, Song L, Zhu P, Zhang B, Tao Y, Xu X, Li F, Wu K, Liang J, Shao D et al (2012) Single-cell exome sequencing and monoclonal evolution of a JAK2-negative myeloproliferative neoplasm. Cell 148:873–885

107. Novak R, Zeng Y, Shuga J, Venugopalan G, Fletcher DA, Smith MT, Mathies RA (2011) Single-cell multiplex gene detection and sequencing with microfluidically generated agarose emulsions. Angew Chem Int Ed Engl 50:390–395

108. Navin N, Kendall J, Troge J, Andrews P, Rodgers L, McIndoo J, Cook K, Stepansky A, Levy D, Esposito D et al (2011) Tumour evolution inferred by single-cell sequencing. Nature 472:90–94

109. Lasken RS (2013) Single-cell sequencing in its prime. Nat Biotechnol 31:211–212

110. Nawy T (2014) Single-cell sequencing. Nat Methods 11:18

111. Panda B (2012) Whither genomic diagnostics tests in India? Indian J Med Paediatr Oncol 33:250–252

112. Xue W, Chen S, Yin H, Tammela T, Papagiannakopoulos T, Joshi NS, Cai W, Yang G, Bronson R, Crowley DG et al (2014) CRISPR-mediated direct mutation of cancer genes in the mouse liver. Nature 514:380–384

113. Sanchez-Rivera FJ, Papagiannakopoulos T, Romero R, Tammela T, Bauer MR, Bhutkar A, Joshi NS, Subbaraj L, Bronson RT, Xue W, Jacks T (2014) Rapid modelling of cooperating genetic events in cancer through somatic genome editing. Nature 516:428–431

114. Matano M, Date S, Shimokawa M, Takano A, Fujii M, Ohta Y, Watanabe T, Kanai T, Sato T (2015) Modeling colorectal cancer using CRISPR-Cas9-mediated engineering of human intestinal organoids. Nat Med 21:256–262

115. Chen S, Sanjana NE, Zheng K, Shalem O, Lee K, Shi X, Scott DA, Song J, Pan JQ, Weissleder R et al (2015) Genome-wide CRISPR screen in a mouse model of tumor growth and metastasis. Cell 160:1246–1260

Printed in the United States
By Bookmasters